EMMANUEL KIZUNGU MULANGANE

soil infertility issues and management

EMMANUEL KIZUNGU MULANGANE

soil infertility issues and management

in agricultural production in the Kagabi group: the case of Mbiza village

ScienciaScripts

Imprint

Cover image: www.ingimage.com

This book is a translation from the original published under ISBN 978-620-6-72420-9.

Publisher:
Sciencia Scripts
is a trademark of
Dodo Books Indian Ocean Ltd. and OmniScriptum S.R.L publishing group

120 High Road, East Finchley, London, N2 9ED, United Kingdom
Str. Armeneasca 28/1, office 1, Chisinau MD-2012, Republic of Moldova, Europe
Printed at: see last page
ISBN: 978-620-8-33020-0

Contents

SUMMARY

In this research on : "The problem of soil fertility and its management in agricultural production in the Kagabi group, specifically in the village of Mbiza". This work raised the problem of soil infertility in MBIZA, which is due to a number of factors, including the use of traditional soil fertility techniques (monoculture, slope cultivation, use of unimproved inputs, etc.), the lack of improved seeds, and the lack of arable land.This being the case, the problem of soil infertility in MBIZA has a negative impact on socio-economic and environmental life. These negative impacts are reflected in a drop in agricultural production, increased poverty and food insecurity, low peasant incomes and environmental degradation.

To overcome these negative impacts and increase agricultural production in the area, the following strategies have been proposed: setting up an extension service for the new improved varieties; introducing an effective agricultural policy; raising farmers' awareness of ISFM; providing technical and economic support for farmers by granting funding and subsidising agriculture.

Key words: Infertility, management, production, agriculture and soil.

GENERAL INTRODUCTION

1.CONTEXT, PROBLEM AND JUSTIFICATION OF THE SUBJECT

Several researchers have focused on the issue of soil fertility problems.

In his book, COOKE (2008) shows that food insecurity hits Africa much harder than other developing regions. The main finding of his research is that most of the total population lives below the $1 a day threshold, 46% of them since the 1980s and 1990s. As a way out of poverty and undernourishment, he proposes the introduction of an agricultural policy to modernise farming and the implementation of a policy of good management of natural resources.

According to the FAO (2015), soil degradation is a real and growing threat caused by unsustainable land use and extreme weather events, which are the result of various social, economic and governance factors. The current rate of land degradation threatens the ability of future generations to meet their needs, unless this trend is reversed through concerted efforts to achieve sustainable land management. The threat to soils compromises the sustainability of agriculture, food security and the provision of ecosystem services.

The FAO recommends that land and soil form the foundation of sustainable agricultural development and that the essential eco-systemic functions of food security be included in nations' sustainable development approaches.

The world's agricultural soils help to feed a growing number of people through the products they produce. Interest in plants has preceded that in the soil. Certain production techniques such as fertilisation have been introduced without any knowledge of soil dynamics (VILAIN, 2000).

Soil degradation is generally a complex phenomenon, involving several elements that contribute to the loss of agricultural potential: soil erosion and removal by water or wind, and loss of fertility resulting from chemical, physical and biological changes (FAO, 2004).

Increasing food security and environmental sustainability in agricultural systems requires an integrated approach to soil fertility management that maximises crop production while minimising the depletion of nutrient reserves and the degradation of soil physical and chemical properties that can lead to land degradation, including soil erosion. Such soil fertility management practices include the use of fertilisers, organic inputs, crop rotation and improved germplasm, combined with the knowledge needed to adopt these practices to local conditions (IAEA, 2013).

However, in many parts of the world, particularly in the tropics, the problems of soil management and soil fertility are acute. Land artificialisation practices, uncontrolled and inadequate urbanisation, poor farming techniques, etc. are all contributing to accelerated soil degradation, with negative consequences for food security and the physical environment (FAO, 2015).

Africa is not immune to these soil fertility management problems. African soils are under threat, with 40% of them suffering from degradation, mainly due to soil erosion, depletion of soil nutrients and loss of soil organic matter (BRABANT, 2000).

Land degradation is a serious problem in Africa. It is reported that more than two-thirds of the production area is affected by land degradation. The economic costs of land mismanagement caused by soil loss and the significant drop in nutrient levels and productivity are estimated at 9 billion dollars per year. More than 3% of agricultural CFM is lost each year as a direct consequence of soil and nutrient loss. Communities suffer the most from the effects of food and energy insecurity and the lack of investment in basic social infrastructure (CGIAS, 2015).

In the Democratic Republic of Congo, soil fertility management problems can be seen in a number of ways: destruction of soil structure, lower organic matter levels, soil leaching, soil acidification, etc. Contracts also show that soils of good productivity account for only around 8% of the total arable area. The majority of land, 90% covering around 42,000 km2 , has average or poor fertility (IITA, 2015). Most of the soils with mediocre and average productivity still have physical characteristics that allow them to be recovered if good soil conservation and soil fertility management techniques are put into practice.

Analyses of the degradation of agricultural land and its impact on agricultural production in the town of Kiliba in the Uvira territory have shown that agricultural yields in Kiliba remain low for the time being, and this is explained by the fact that 100% of the samples claim that their production has become low as a result of the infertility of the soil. This is explained by the fact that 70% of the surveys claim that the soil is less fertile, especially as a result of poor cultivation techniques in Kiliba. After noting the problems in Kiliba, the author suggested some possible solutions that would enable farmers to increase agricultural production by using new cultivation techniques and modernising agriculture, which are strategies that should be implemented to increase agricultural production (KIFUVWE MAKI, 2013).

In her research on soil degradation and its impact on agricultural production in Lubero, GATUMBA J. (2015) showed that ignorance of soil conservation techniques is one of the factors preventing the population in her study area from increasing agricultural production. She showed that erosion and poor cultivation techniques degrade the soil, crops become meagre and food becomes insufficient; all these problems affect aspects of the inadequacy of cultivated land. For this reason, the improvement of new cultivation techniques would be a strategy for revitalising agriculture in Lubero.

Still in the context of soil fertility management, it has been shown that improving the fallow period by introducing leguminous plants sown in association under the last crop after weeding will facilitate the reconstitution of fertilising elements, while resting plots and using them for grazing small ruminant livestock will enable the manure left by the animals to be put to good use. He completes his ideas by proposing an improvement in breeding methods for small livestock, which will consist of rationally moving the stakes in order to avoid over-exploitation of the same sectors. For large livestock, it is necessary to institute a semi-stable system with a short transhumance during the dry season over the whole of the pastoral territory, and the construction of compost bins is proposed. The author points out that in inhabited areas, this practice could be generalised, starting with the holes from which the earth for the bricks is extracted, the pulp and poles from the coffee, the dregs from the brewery, the residues from the soilless livestock, the banana stalks will be piled up and will decompose slowly in the shade of the trees in the vicinity of the inhabitants; Slowly humifying waste will be left on the spot, while part of the

generalisation of market gardening, the final solution it has found is solely the creation of fertiliser banks in relay districts between farmers' associations and agricultural cooperatives. The latter are a fundamental necessity for any sustainable project (FOTSING, 2002).
Soil tillage covers all the cultural activities that create or maintain a structural state of the soil that is favourable to plant development by facilitating germination, growth and root establishment. It also allows the burial of harvest residues, soil improvers, mineral and organic fertilisers and the destruction of harmful vegetation. These means of action available to agriculture are completed by the influence of atmospheric agents, their effects and the result of alternating frosts and thaws, wetting and drying (DETRAUX F. and OESTGES, 2000).
In our study, which was carried out in South Kivu in the KABARE territory, in the KAGABI groupement, in the MBIZA village to be precise, it was revealed that for several decades, agriculture in all the territories of South Kivu has been faced with several problems, including a reduction in agricultural production, the appearance of new diseases causing the disappearance of certain key crops, the problem of insufficient arable land and infertility of the soil.
In the Kagabi group, which is the subject of our research, fallow land, a natural means of restoring soil fertility, has become almost non-existent. The soil's fertility is being depleted year after year, and the resources deployed to deal with the problem are minimal compared with the seriousness of the problem.
According to observations made on soil management in the Kagabi groupement, this is not effective and is justified by the fact that agricultural production is still low despite the fact that farmers here use certain organic fertilisers. This low level of soil production in MBIZA is a major problem, given that the use of organic fertilisers has a negative impact because production remains low.
In view of this sad situation, we would like to ask the following questions:.

> What are the factors contributing to soil infertility in the village of MBIZA?
> What are the impacts of soil infertility on socio-economic and environmental life in the village of MBIZA?
> What strategies should be put in place for good soil fertility management and increased agricultural production in the village of MBIZA?

2 .WORK HYPOTHESES

Based on the questions posed above, we offer the following provisional answers:
S The factors contributing to soil infertility in the MBIZA village can be observed in traditional agriculture.
S Soil infertility is said to have a negative impact on the socio-economic life of farmers in the village of MBIZA.
> Rational management of soil fertility to increase agricultural production in MBIZA Village would require the introduction of strategies to raise awareness of the need to adhere to the agricultural calendar and to provide farmers with technical and economic support.

3 . OBJECTIVES OF THE WORK

OVERALL OBJECTIVE

The aim of this work is to analyse the problems of soil infertility in the Kagabi group, specifically in the MBIZA village, and how they are managed in agricultural production.

SPECIFIC OBJECTIVES

> Identify the factors of soil infertility observed in the MBIZA village,

> Determine the impact on the socio-economic life of the population of MBIZA Village.

> To identify strategies for good soil fertility management and increased agricultural production in the village.

Object of analysis

Our aim would be to analyse soil fertility management methods and productivity (cultivation techniques, fertilisers used, farmers' knowledge of soil management, etc.) in the village of MBIZA.

4 .SPACE-TIME DELIMITATION

We are delimiting this research in space and time for greater precision.

❖ **Spatially**: we are seeking to analyse the problems of soil infertility and its management in agricultural production in the Kagabi group and particularly in the MBIZA village.

❖ **In time**: our research covers the 6-month period from the beginning of March 2024 to the beginning of September 2024. This period corresponds to our analyses of the issues addressed in the MBIZA Village.

5 .CHOICE AND INTEREST OF SUBJECT

The choice of this subject is to know the reality on the ground by observing the texture of the soil and the mode of management of the soil by the farmers in order to determine the causes of the infertility of the Soil in the village MBIZA and its impact on the socio-economic life.

- From a scientific point of view, to have the data available so that other people approaching this area of research can find the documentation and other information they need to guide their research.
- On a personal level, this work is a management of the theories learned at the auditorium and the practices carried out in the field, as well as acquiring more knowledge of the environment in relation to soil management.
- On a social level, this work is an appeal to the entire urban population and its surroundings, in particular that of the village of MBIZA, to implement the various solutions proposed in this work with a view to putting an end to the phenomena of soil infertility in the village of MBIZA and other environmental indicators.

6 . DIFFICULTIES ENCOUNTERED

Writing a memoir is a human endeavour, requiring bravery and countless sacrifices. This being the case, throughout our scientific work, we have come up against a number of constraints, including :

- The lack of access to the works expected to improve the writing of this work is due to the impasse reached by some librarians at Buka vienne institutions and universities.
- Financial resources were a major challenge for us throughout our investigations.

- Lack of adequate material resources, such as the accessibility of the road in the village of MBIZA due to the lack of rolling stakes for carrying out this work.

7 .SUMMARY SUBDIVISION OF WORK

Apart from the introduction and general conclusion, this work is divided into three chapters:

The first chapter deals with the Theoretical, Conceptual and Methodological framework of the work.

The second chapter deals with the study environment and the theoretical aspects of the work.

The third chapter deals with the presentation, analysis, interpretation and discussion of the results.

CHAPTER I

REVIEW OF LITERATURE

I.1. CONCEPTUAL FRAMEWORK

It is important to clarify the key concepts of the study so that they can be understood by our readers, depending on the context in which they are used and the strategies adapted.

1. The floor

Soil is an essential component of the environment: "it is a critical zone in contact with the atmosphere within which interactions occur with water, air and living organisms" (TESSIER et al. 2002).

Soil provides essential ecological services. The first service it provides to mankind is primary agricultural production, whether for food (intended for humans or livestock) or non-food uses (agrofuels, fibre production for textiles). Soil also provides environmental services related to air, water and biodiversity. Above all, it is a living environment that is home to a large number of organisms that have a direct impact on soil fertility in relation to air.

2. Soil fertility

Soil fertility is a very important concept in agriculture and agronomy, designating a soil's ability to produce under current growing conditions. It is one of the components of soil quality (FAO, 2015).

Soil fertility refers to the soil's ability to support plant growth and optimise crop yields. It can be improved by applying organic and inorganic fertilisers to the soil. Nuclear techniques provide data that can be used to improve soil fertility and agricultural production while minimising environmental consequences (FA0, 2006).

According to SEBILLOTTE (2010), soil fertility is its capacity to meet the physical, chemical and biological requirements necessary for plant growth, productivity, reproduction and quality (considered in terms of human and animal well-being in the case of plants used as food or fodder), in a way that is adapted to the type of plant, soil type, land use and climatic conditions.

3. Soil infertility

Generally speaking, soil infertility refers to the loss of a soil's buffering capacity, i.e. its inability to produce the maximum yield required under normal conditions. Soil infertility can occur as a result of excessive ploughing, which damages the soil structure. According to experts, an accumulation of salt in the soil, often associated with irrigation, can dry up fertility and limit crop yields (FAO, 2015).

4. Agricultural production

Agricultural production refers to all the activities developed by man with the aim of transforming his natural environment in order to produce plants and animals that are useful to him, in particular those aimed at obtaining from the soil as many of the necessary products indispensable for human and plant consumption as possible.

I .2 .THEORETICAL FRAMEWORK

A. Theory of Integrated Soil Fertility Management (ISFM)

ISFM refers to the set of farming practices adapted to local conditions to maximise nutrient and water use efficiency and improve agricultural productivity. The approach includes all the resources available in the integrated soil fertility management programme: fertilisers, locally available soil conditioners, organic matter such as crop residues, compost, green manure and intercropping (IFDC, 2015).

The GIFS approach is based on the following key principles:

> Neither practices based solely on mineral fertilisers nor those based solely on the management of organic matter are sufficient to ensure sustainable agricultural production;

> Perfectly adapted genetic material that is resistant to disease and pests is needed to make efficient use of the nutrients available.

> Good agronomic practices in terms of planting dates, planting density and weed control are essential to ensure efficient use of scarce nutrient resources.

In addition to these principles, integrated soil fertility management recognises the need to target nutrient resources in crop rotation cycles, preferably by integrating legumes, thus going beyond recommendations in favour of pure crops. Successful adoption of ISFM significantly improves productivity. In addition, there is a positive synergistic effect between organic and inorganic inputs. ISFM recommends strategic choices regarding the timing and placement of fertiliser administration using inorganic nitrogen fertilisers, often at significantly lower rates than recommendations based on the use of inorganic fertilisers alone. (SANGINGA et al. 2009).

On the basis of this ISFM theory, this work is designed to circumvent any problems associated with the poor management of agricultural soils currently prevalent in MBIZA Village, such as overexploitation of the soil and the preponderance of human activities (mining, overpopulation, etc.), which cause infertility in the soil. This would encourage farmers to respect the GIZ principles in order to increase agricultural production in their areas and improve their socio-economic conditions.

B. Description of strategies

STRATEGY 1: INVOLVING THE AUTHORITIES

This strategy will make it possible to reduce overexploitation of the soil and, above all, to strengthen the soil conservation process by restoring soil throughout the KABARE area, particularly in the village of MBIZA. This will make it possible to improve the agroecological conditions in the area, which will help to improve agricultural yields.

STRATEGY 2: RAISING AWARENESS OF GIFS

This strategy will enable the village community to take responsibility for human actions harmful to the soil, in order to change the way it manages the soil and guarantee food security in its environment. The relevant services will be mobilised to train pilot farmers in the key principles of integrated soil fertility management. After the training, the population will be provided with the necessary tools to enable them to make better use of the agricultural soils in their area.

STRATEGY 3: SETTING UP AN EXTENSION SERVICE FOR NEW CULTIVATION TECHNIQUES

The role of these services will be to raise public awareness of the importance of rational management and conservation of agricultural soils. These services will also have the task of popularising new cultivation techniques to increase agricultural yields.

STRATEGY 4: INTRODUCING A BETTER AGRICULTURAL POLICY

This will enable agriculture to be modernised and natural resources to be properly managed. It is up to the Congolese government to introduce this policy, as agriculture is one of the priority development sectors that will help secure the country's food supply. As agriculture is underpinned in the DRC in general and in MBIZA in particular, it is important to consider a better agricultural policy that will increase agricultural yields in the area.

STRATEGY 5: FACILITATE FARMERS' ACCESS TO AGRICULTURAL FINANCE

It seems that many farmers in rural areas of the Congo are struggling to carry out their agricultural activities because they do not have access to agricultural credit. This being the case, this strategy seems vital in that it will enable farmers in the KABARE area to benefit from funding from NGOs to relaunch agricultural activity and improve their socio-economic conditions.

C. Roles of the players and their interactions

1. The State

The role of the State is to reform the agricultural sector and protect natural resources. It must introduce a policy that will enable the other players to work well, such as encouraging and supporting the activities of NGOs working in the agricultural sector, in order to revive agricultural activity. It must mobilise all departments to work in synergy to ensure good management of agricultural soil fertility in order to increase agricultural production.

2. The population

It will be called upon to apply the cultivation techniques that will be made available to it by the competent services. They must behave responsibly towards the soil and the human activities that cause soil degradation.

The population, as direct beneficiaries, will be involved in all phases of the strategy's implementation. These beneficiaries will have to continue basic actions in the fight against soil infertility in their environment through training tools and information on natural resource management. This will only be possible if the population forgets the bad mentality of practising bad cultivation techniques and uses the best.

3. The experts

The team of trainers must make sure that the agricultural and environmental service agents have fully understood the advice and that they have covered all the appropriate points for reconstituting the soil in all the villages.

Private investors (NGOs, associations) working in these areas should provide financial support to the agriculture and environment departments and carry out regular monitoring and evaluation to check on the progress of soil restoration work by the local population and the application of farming techniques that have been popularised.

4. Extension services for new cultivation techniques

These services must make efforts to disseminate new farming techniques in order to improve agricultural yields, which are currently low as a result of the many factors elucidated in the preceding pages that are observed in the KAGABI group in general and in the MBIZA village in particular. These services must ensure that they have the necessary tools for restoring agricultural soils and the necessary means for reconstituting soils and popularising cultivation techniques in the area.

D. Viability of strategies

The viability of our strategies depends on the effective participation and support of all those involved in the planning process. Consequently, the population must have access to the information, standards and cultivation techniques that the services will be able to put in place to fertilise the soil. This would make it possible to combat any problems resulting from the infertility of the land in the area.

We therefore believe that our strategy is viable, as it will help to further combat soil infertility in the KAGABI group, and MBIZA in particular.

E .Activities

To combat soil infertility in our field of study, we are going to undertake a number of activities to mobilise the population to tackle this environmental problem, which has a negative impact on the socio-economic life of farmers.

Given that the success of our strategy to promote good soil fertility management and increase agricultural production requires the support of all stakeholders, here are the main activities to be carried out:

- Raising public awareness of the importance of good farming techniques in the daily lives of farmers, to help them achieve self-domination;
- Train local facilitators in ISFM;
- Restoring nutrient-depleted agricultural soils;
- Raising awareness of environmental management ;
- Training local facilitators in the soil management process
- Intensifying agroforestry in the fields.

F. Strategy objectives

For the objectives, being considered as the whole to be achieved. These are necessary in the sense that improving the agricultural and socio-economic soil conditions of farmers in different villages and thereby contributing to the country's development will be an asset in the fight against soil degradation. To increase the availability of fertilisers and improved seeds throughout the villages; to disseminate new cultivation techniques in order to set up village composts to stimulate the interest and establishment of stakeholders; **to** step up environmental and agricultural education in the villages.

G .Strategy indicators

The indicators of the responsible land use strategy are as follows:

- Anthropogenic pressure on agricultural soils is decreasing by 90%;
- 80% of good cultivation techniques popularised;
- Soil infertility 85% under control
- Floors restored to 75%;
- The socio-economic conditions of farmers improved by 30 to 75%.

THEORETICAL ASPECT OF THE WORK

b. The principles of nature conservation to combat soil infertility.

Wiegleb (2002) defines eleven guiding principles of nature conservation:

- Prevention of environmental poisoning: prevention of toxic emissions. Monitoring the quantities of toxic substances released into the environment.
- Protecting ecological processes: guaranteeing natural processes and phenomena (evolution, succession, soil formation).
- Nature conservation: conserving the original elements of the landscape, i.e. intact elements of nature.
- Minimising the intensity of land use minimising disturbance and anthropogenic influences.
- Conservation of what is authentic: geological and geomorphological formations, biological adaptations (geysers, rare rock formations, endemic species, rarity in the context of a region, a country, a continent, the world).
- Preserving beauty: aesthetic landscapes, views, attractive features (mountain lakes, rocks, trees, special animals and plants).
- Aiming for symbiosis with nature: species that help or are helped by the development of others.
- Mystical and religious attitudes to nature: sacred high places, natural formations associated with the history of the gods, fairytales, esotericism, etc.
- Achieving and maintaining sustainability: protection and rehabilitation of landscape functions or natural capital to promote sustainable crop production, soil protection (prevention of erosion and groundwater pollution).
- Conservation of diversity at species level optimising or maximising the number of species, protecting key species (umbrella species).
- Conservation of diversity in landscapes different types of habitat, diverse landscapes, mosaics.

The 11 guiding principles of nature conservation can be grouped into more comprehensive categories and are complementary.

The prevention of environmental poisoning, the conservation of ecological processes, the conservation of wildlife and the minimisation of land use intensity, impact and disturbance have a strong protectionist connotation. There is a strong reference to what is natural or close to natural.

The conservation of what is authentic, and approaches to spirituality and beauty, come under the protection of cultural landscapes or cultural values. The protection of biodiversity is divided into direct approaches (based on the protection of species and other biotic elements) and indirect approaches (based on substitutes such as habitats and landscapes).

- The new scientifically-oriented concepts of sustainability and biodiversity protection have a strong functionalist connotation; they are linked by the concepts of biotic resources and ecological services.

The focus on sustainability since the publication of the Brundtland Report and the Rio Conference has made sustainable development an obligation in terms of resource management. A multi-dimensional concept, it integrates economic, environmental and social concerns. Sustainable development gives rise to a multitude of interpretations, which in turn give rise to several approaches to natural resource management. These can be distinguished schematically according to whether they are based on a strong or weak

sustainability principle, or even on the precautionary principle (Debailleul, 1909).
The principle of weak sustainability is based on natural resource management rules which postulate that natural capital can be entirely replaced by different forms of reproducible capital or by different forms of technical progress. According to this approach, a loss of soil due to erosion can be fully compensated for by more efficient use of inputs or by technical innovations. In our poor countries and tropical climate, the application of this principle can lead to the irreversible destruction of natural resources. Tropical ecosystems are very fragile (Maldague, 1997) In terms of the principle of strong sustainability, if we have doubts about the degree of substitution between different categories of capital, about the rate of technical progress, if we take into account the difference between the situation of natural capital and technical capital in the face of irreversibility (existence of a critical zone under conditions of reproduction of natural capital), we will adopt more "conservative" management principles with regard to natural resources.
The precautionary principle adds to the postulates of strong sustainability, risk aversion and concern for the protection of future generations.

c. Soil conservation.

It is essential to use the soil correctly to satisfy immediate needs, to protect it for future use and to maintain fertility and water resources. To achieve this, erosion must be combated and salinisation minimised.

c.1. Combating soil erosion

Action should be taken to reduce the effects of rain on the soil, by reducing the kinetic energy by interposing a plant screen on the soil. It is also possible to increase soil resistance by planting vegetation. Steep slopes and soils that are highly susceptible to erosion should be left under permanent cover: forest, bush or savannah. Plant cover prevents erosion in several ways. It protects the soil from the impact of raindrops, increases the capacity of water to infiltrate the soil and reduces the amount of water that runs off the surface. It reduces the speed at which surface water runs off. Through the action of plant roots, plant cover helps to hold the soil in place. To reduce run-off where it exists, the slope can be reduced by appropriate cultivation methods or the speed of run-off can be reduced by civil engineering works. Depending on their mode of action, erosion control methods can be classified as follows:

- biological processes, giving soils greater resistance to water attack through vegetation grown using so-called "conservative" methods.
- Mechanical processes to control runoff and, consequently, erosion, by reducing or eliminating the impact of the slope.

<u>Agroforestry, one of the biological precedents</u>

Agroforestry refers to the set of land use systems where perennial woody plants are deliberately associated with crops or livestock production in the form of a spatial arrangement or temporal sequence taking place on a single land management unit. Agroforestry systems are characterised by significant ecological and economic interactions between their various components (Lundgren and Raintree, 1982; cited by Olivier, 1997).
All agroforestry systems share the following characteristics: they involve at least two

associated species, at least one of which is woody; there are at least two products from the system; and the production cycle extends over more than one year. The concept of agroforestry can be divided into 3 main areas: agri-silviculture, sylvo-pastoral integration and agro-silvo-pastoral integration.

Agroforestry systems for soil conservation generally involve planting trees along contour lines and at the edges of agricultural plots to help combat water erosion.

Agroforestry has an influence on soils. One of the main advantages of agroforestry is its ability to positively modify the physical properties of the soil. This is important because soil fertility conservation is the basis for the sustainability of agricultural and forestry production systems. Maintaining the rate of increase of organic matter in the soil, improving the cycle of mineral elements, nitrogen fixation, improving the physical properties of the soil and protecting it from water and wind erosion are the main beneficial effects of agroforestry on soils.

Agroforestry has ecological and economic advantages. One of the significant advantages of agroforestry over monocultures and mono-specific plantations is its ability to increase plant and animal diversity. Thus, in coffee plantations under tree cover, biodiversity is much higher than in industrial plantations without shade trees (Olivier, 1997).

From an economic point of view, many of the trees and shrubs used in agroforestry, as well as certain herbaceous plants, are leguminous plants that carry out symbiotic fixation of atmospheric nitrogen by favouring the accumulation of leaf litter or the application to the soil of the progeny of nitrogen-fixing woody species. Certain agroforestry systems enable the nitrogen content of the soil to be enriched. For example, by combining maize with nitrogen-fixing shrubs, the topsprouts of which are applied to the soil, the cereal is provided with a rapidly available source of nitrogen, thereby improving its yield.

C. Some techniques for combating run-off

- Living hedges

Plant hedges or create ridges on contour lines. A contour line is a line that connects all points on a topographic surface that are at the same height. It can be drawn using a simple tool called a slope triangle. The ridges on the contour lines are formed by herbaceous species such as Tripsacum, Setaria, etc. Bulges are heaps of leaves or stones piled up on the contour lines which break the vertical flow of water down a slope and prevent the formation of gullies.

- Crop rotation

Crop rotation involves dividing the land of a farm into plots: one plot for each crop. This plot is called a sole. Crops must change plots after each season. When the same crop is grown over and over again on the same plot of land, the land becomes poorer and poorer. For example, groundnuts have small roots, while millet has deep roots. These plants will therefore exploit different layers of the soil. They also have different requirements for mineral nutrients: millet needs nitrogen, tuberous plants need a lot of potash and forage crops need a lot of phosphorus.

Rotation consists of moving a crop from one plot to another (and vice versa) over a period of time, and in succession. Within the same plot, you can alternate weed crops that encourage erosion (maize, rainfed rice, sorghum, millet, groundnuts, cotton, etc.) with cover crops or green manure plants.

Crop rotation encourages biological control of crop diseases, especially when successive plants are not attacked by the same species of insect pest. By encouraging biological control of crop diseases, we reduce the harmful effects of pesticides on the environment.

- The construction of engineering structures

In a region affected by landslides, such as Bukavu, erosion control may require the construction of engineering structures (Mweze, 2003),

D . Combating soil salinisation

Some plants have a natural tolerance to salt: date palms, alfalfa, olive trees (Robert, 1996). We can also try to adapt species or transform them to adapt to salt or water stress, which are similar in nature. It should be borne in mind that there is a certain level of salinity above which vegetation can no longer adapt, giving way to desertification. Irrigation management is a delicate technique that needs to be implemented: irrigating with water with a low salt content. In our country, research should be undertaken to identify salt-tolerant plants.

E .Amenagement

Soil conservation measures such as erosion control, agroforestry and the adoption of improved farming techniques all contribute to water protection. Other measures, such as eutrophication of banks and riverbanks, watershed management and control will be dealt with under this heading.

1. Stabilising banks and riverbanks

It has been observed in the field and in the laboratory that current velocity is at its highest slightly downstream of the axes of curvature of the meanders (Clibbon, 1997). Erosion will therefore be particularly significant in these areas. Mechanical protection techniques consist of building a concrete or gabion retaining wall along the concave bank. A bank can also be stabilised with stone riprap. The bank can also be stabilised by planting vegetation. This technique is both ecologically sound and inexpensive. As grass is not enough to hold the soil in place, trees and shrubs with dense, well-developed root systems must be used. Establishing a strip of riparian vegetation along a watercourse is also effective. The width of the riparian strip varies between 10 and 50 m in most cases (Jorgensen and Loffer, 1990). A strip of riparian, herbaceous, shrubby or wooded vegetation protects the banks from erosion caused by water running off the fields, provides shelter and food for wildlife, provides shaded areas and keeps the water cool. Wooded riparian strips are home to a variety of fauna, some of which are useful to the farm, such as certain birds that can destroy insects harmful to crops. The woodlands that develop along watercourses are an important source of plant debris, fruit and insects that fall into the water and are eaten by invertebrates and fish.

2. Intervention at catchment scale

The area of land drained by a main river and its tributaries is called a watershed. From the upper end to the base of a catchment, water moves and flows into larger and larger streams to reach the main river or water body. The natural characteristics of a catchment and the activities carried out in the area play a decisive role in determining water quality. If a catchment is used for agriculture, river water will be affected by erosion or pollution from fertilisers. If there is little human activity in a wooded catchment area, the rivers and water bodies will be less polluted in the absence of acid rain and disruptive natural phenomena. Catchment management is an aspect of integrated land-use planning. Two categories of measures can be distinguished (Maldague, 1997):

a. management measures are aimed at (very frequent cases) In this case, in order to maintain the natural balance, it is necessary to ensure the permanence of as complete a plant cover, forest or permanent grassland, as possible. For example, in the case of steep, rocky terrain, the forest must be preserved. This is a preventive measure.

b. However, it may happen that corrective measures (afforestation, for example) need to be considered. In this case, the fight against degradation phenomena (erosion, torrential erosion, for example) must be undertaken from the top down rather than from the bottom up. The correction of the catchment area must precede the development of the land located downstream.

3. Combating eutrophication

The fight against eutrophication and water poisoning essentially involves draining all sources of waste water and purifying the water supplied to rivers and lakes. The use of properly designed septic tanks, with the water discharged into absorbent wells, is effective. The use of chemical fertilisers should not be exaggerated. Organic fertilisers should be used sparingly. Regular monitoring of water quality provides the data needed to make timely decisions.

4. Conservation of flora and fauna

To conserve flora and fauna, the following measures are recommended: the fight against the fragmentation of ecosystems and biological invasions, the fight against bush fires, management by star species and the designation of portions of the national area as protected areas for species representative of the region's flora and fauna.

4.1. Combating ecosystem fragmentation

Deforestation does not necessarily result in the complete disappearance of the forest cover, but often in its fragmentation into patches of varying size. Fragmentation has a considerable impact on wildlife, as it deprives them of communication routes between forest patches. However, it is possible to maintain communication routes between plots by preserving natural vegetation corridors. Corridors are at the centre of a controversy: some believe that corridors expose wildlife to danger (trapping or easy hunting in a small area); others argue that corridors encourage exchanges between separate plots. Belisle found that mesangutans (birds) used wooded corridors more than open areas (Lavoie, 1998) to move from one plot to another. In the Kahuzi-Biega National Park in the Democratic Republic of Congo, Refisch (1991) observed that the corridor linking the high-altitude part (1,800 to 3,308 m) to the low-altitude part (600 to 1,200 m) favoured exchanges between animals from these two environments. This exchange between animals in Kahuzi-Biega is seen as a means of preventing inbreeding in gorilla populations confined to the 600 km^2 high altitude part of the Park (Mulhenberg et al., 1995).

4.2. The fight against biological invasions

By definition, a biological invasion is the geographical expansion of species into an area they did not occupy before. Invasions can occur as a result of climate change or through the introduction of species by humans.

To avoid the damage caused by biological invasions, a biodiversity monitoring system needs to be put in place. With this system, it is possible to know which species have been preserved and to notice in good time when a new species appears in the ecosystem.

In-depth research must therefore be carried out to determine the consequences of introducing a new species into a given environment. For example, the request to introduce the River Barbus ictalurus punctatus, a sub-species of omnivorous fish, was rejected by the New Zealand government on the basis of the results of a study which showed that this fish would have a negative impact on the survival of at least four species of indigenous fish (Leveque, 1994).

4.3. The fight against bush fires

When an area of forest which has already been degraded (for example by slash-and-burn agriculture) is fully protected against bush fires, a secondary forest is formed, tending towards the formation of a dense semi-deciduous forest (Ministry of Cooperation, 1979). Integral protection is necessarily limited for economic reasons, but it must be extended to all forest areas where conservation is necessary. The creation of firebreaks at the end of the rainy season is a practice that makes it possible to prevent fire in protected forests.

Among certain African herders and hunters, the practice of burning is often an ancestral habit. If it is difficult to prohibit fires, early fires can be recommended. Early fires are practised at the beginning of the dry season, when the leaves on the trees and grasses are still green. At this time, water from rain is still present in the vegetation. The practice of late fires should be banned. In this case, tighten up the regulations on burning.

4.4. Development by star species

By selecting a single species to take centre stage, an entire area can be preserved with all its biodiversity. The star species acts as an umbrella for the other animal and plant species in the ecosystem. In the Kahuzi-Biega National Park, the gorilla is the umbrella animal. By protecting it, we protect its habitat and the other species that live there.

There are two aims to the management of star species:

- To encourage the development of a given species by ensuring the sustained productivity of its habitat;
-Simultaneous management of wildlife and vegetation, especially woody vegetation.

Management by star animal is based on the following principle: once a single species of interest has been selected for a given area, the management of this area is adapted to meet the habitat requirements of the species.

Some selection criteria for the star species are :

- Public interest in the species ;
- Threatened species ;
- Complementarity between species production and wood production.

f. Using and enhancing biodiversity

Biodiversity is the basis of all our food resources and has many applications in the fields of health, industry and biotechnology. Today's domesticated animal and plant species are of wild origin. Plants that are particularly important in the daily lives of the Aztecs of Mexico

Pre-Columbian crops are now grown all over the world, including maize, beans, sweet potatoes, squash, cocoa, cotton, tobacco and chilli.
The domestication process continues with the search for new breeds or varieties that perform better, with better growth and/or higher yields. For example, to combat coffee tree diseases that are resistant to the use of fungicides, genetic strains from wild populations in the coffee's centres of origin - Ethiopia, the high plateaux of East Africa and the rainforests of Madagascar - are used.
A number of manufacturers have undertaken research to make better use of the active ingredients in medicinal plants. The Pharmakina company is currently popularising *Prunus africana* among farmers. Good Prunus africana seeds have been collected in the Kahuzi-Biega National Park. Extracts from this plant are used to make medicines to treat prostate cancer. The use of medicinal plants is a widespread and ancient phenomenon. Almost 80% of humanity has no access to medicines and continues to rely on the virtues of plants to treat itself.
Animals are also the source of pharmacological substances. For example, shark liver contains substances that increase the human body's resistance to cancerous diseases, while bee venom is used to treat arthritis.
Nature provides a reservoir of industrial products: timber, wood for paper pulp, rubber, oils extracted from trees or animals, etc.

G. Importance of natural resources in combating soil infertility

A natural resource is a good, substance or object found in nature and exploited for the needs of a human society. Natural resources include water, air, earth and soil, as well as raw materials (metals, ores), living organisms (micro-organisms, plant matter, forests, animals) and fossil fuels (coal, oil, natural gas). We often think of natural resources as sources of energy, whereas they are first and foremost conditions for the emergence of life in general (the sun, water, air, living species).
Natural resources can be used in their raw state, albeit using a variety of processes that do not alter them (this is the case for plants and animals, but also for renewable energies derived from the air, wind, water and sun). They can also be transformed in order to be exploited, and these are essentially fossil energies such as coal, oil, natural gas or uranium.
Since the 1970s, this concept has evolved to include resources useful to all ecosystems and all socio-economic sectors. Soil availability, water or air quality, the appearance of landscapes and biodiversity are other aspects of natural resources.
The fragility and limitations of certain resources characterise non-renewable resources (e.g. speech), as opposed to renewable resources (e.g. biomass), which are not inexhaustible.
Sensible management of natural resources is an essential condition for human existence. The question of natural resources is absolutely essential when it comes to sustainable development; it is even its foundation, since the doctrine of sustainable development emerged from the notion of risk, which arose when societies began to question their relationship with nature and its resources.
Talking about natural resources is therefore tantamount to considering nature as providing Thomas with the means to ensure his development. We must no doubt always bear in

mind and take note of the fact that any discourse including natural resources stems from a utilitarian vision of the world - in general, it is a question of determining at what rate and under what conditions such and such a resource can be exploited. From this point of view, any discourse that talks about natural resources while, for example, prohibiting their use could be considered contradictory.

The concept of nature conservation encompasses two dimensions: integral protection (static aspect) and rational management (dynamic aspect).

Conservation Integral protection Rational development

Rational development allows natural resources to be exploited in the best possible way, which means ensuring that the initial capital is preserved and used sustainably. T he management of natural resources concerns the processes and practices involved in their allocation and use. Sustainable management makes the best possible use of natural resources to meet current subsistence needs, while maintaining and improving the quality of natural resources for future generations.

Nature conservation extends the scope of its concerns not only to natural resources, but also to natural and modified environments, as well as to the mechanisms by which nature is produced.

Resources (for example, the cycles of biogenic elements through which natural resources are regenerated) For Nicholson:

"Nature conservation is everything that mankind plans and does to soften his impact on his natural environment and to satisfy all his *true* needs*, while allowing this environment to perpetuate itself in a state of balance and functioning.*

The aim of nature conservation is to find ways of balancing resources. With the revolution in society, the concept of nature conservation, while presenting a deliberately interventionist aspect, is still too limited in its scope, since it essentially targets natural resources and natural ecosystems.

Natural resources and the services they provide are essential for food production, strengthening rural development and sustaining rural livelihoods.

Conflicts and competition over access to and use of these resources are likely to increase in many regions, due to the sharp rise in demand for food, fibre and energy, and the loss and degradation of productive land. These conflicts will be exacerbated by changing cropping conditions, increased biodiversity loss, extreme weather events and other effects of climate change.

The sustainable management of natural resources for present and future generations requires that certain key aspects of natural resources and their governance be addressed today, across all sectors. As underlined by the Millennium Ecosystem Assessment report, published in 2005, no progress in addressing the Millennium Development Goals of poverty eradication, poverty reduction and environmental sustainability can be achieved if human ecosystem services continue to be degraded.

Many natural resources are concentrated in a small number of countries with limited reserves. For example, statistical data shows that 90 per cent of the world's proven oil reserves are found in just 15 countries (out of 200 countries in the world today), and 99 per cent of the world's oil reserves are found in 40 countries. International trade can help to remedy these disparities in natural resource endowments by enabling the transfer of

resources from regions where supply is low to regions where demand is high, which can also contribute to their more efficient use.
However, as natural resources are essential inputs for production, they are also necessary for maintaining people's quality of life, and their unequal distribution can be a source of tension between nations.
In most trade disputes involving agricultural products or manufactures, a country seeks to limit imports for a variety of reasons - for budgetary reasons, to support a nascent or "strategic" industry, for reasons of public interest (health, environment, safety, etc.), or in response to trade practices that the importing country considers unfair. Conversely, most countries import natural resources from foreign suppliers. But exporting countries may be reluctant to let their resources go to other countries for a variety of reasons, including to meet budget requirements, to diversify the economy by developing the processing of raw materials, to guarantee adequate domestic supplies and to protect the environment.
The uneven geographical distribution of traded natural resources is also illustrated by Maps 1 to 5 in the Appendix, which show net exporters and net importers by commodity, based on merchandise trade data from the UN Contra database. Of particular interest is the distribution of fuels and non-ferrous metals, all the major industrial economies being net importers of these products. All European countries, with a few exceptions, are net importers of all types of natural resources, as are Japan and the Republic of Korea. The United States is a net exporter of forest products and minerals, but a net importer of all other tradable resources. India and China are only net exporters of fish, but net importers of many other resources. Russia is a net exporter of all resources except fish. Among the major developed economies, only Canada is a net exporter of almost all types of natural resources.
Water, which is not generally traded, is also very unevenly distributed between countries. According to the United Nations, humanity is facing a serious water scarcity problem (United Nations, 2009). Most of the world's water resources are salt water, with freshwater accounting for only 2.5 per cent. Approximately 70 per cent of the available freshwater is found in the ice caps of Antarctica and Greenland, leaving only 17 per cent of the world's water resources for consumption, and of this 0.7 per cent, approximately one per cent is used for agriculture. The world's finite reserves of drinking water fit for human consumption are finite and rapidly diminishing, posing a serious threat to public health, political stability and the environment.
Water scarcity is exacerbated by population growth, increasing urbanisation and high levels of per capita consumption, among other factors. Climate change is also expected to contribute to water scarcity in the future, as rising temperatures will lead to droughts, desertification and increased demand for water. The problem of water scarcity is more acute in some countries than in others, as the per capita supply of water is much higher than in others. Water itself cannot be traded, and international trade can have a much greater effect in countries such as Canada, Russia and Brazil than in the Middle East and large parts of Africa. For example, in Canada, it is 87,000 m^2 per person per year, which is around nine times higher than in the United States (9,800 m2 per year), where it is almost 14 times higher than in Egypt (700 m2 per person per year). And in Egypt, the water supply is around seven times greater than in Saudi Arabia, where resources are only 95 m

per person per year (Food and Agriculture Organization of the United Nations, AQUASTAT database). International trade could possibly help to alleviate water scarcity problems at local level by providing resources where they are most needed. But countries are unable or unwilling to do this because there are virtually no large-scale shipments. This lack of trade is largely due to technical factors, as water is a heavy commodity and therefore difficult to transport. Furthermore, water scarcity or abundance is generally shared by most countries in the same region, so water would have to be transported over long distances to overcome the scarcity problem. Although indirectly beneficial to countries' water supplies. Exporting water-intensive products (such as agricultural products) from regions where water is abundant to those where it is scarce can enable importing countries to save money by freeing up resources for other uses. For example, between 1997 and 2001, importing water-intensive goods into Japan saved the country 94 billion m' of water that would have been needed to produce these goods domestically (Hoekstra, 2009).

H. Dematerialisation of the economy and conservation of natural resources

In 1992, Chapter 4 of Rio's Agenda 21 was devoted to the need to change consumption patterns, by making better use of energy and resources, reducing waste production to a minimum, and gearing the choices of individuals and actors towards ecologically sound products and practices.

5 years later, in 1997, the General Assembly of the United Nations clarified this orientation by associating it with quantified objectives formulated in these terms: "studies proposing a more rational use of resources should be examined and consideration should be given in particular to multiplying by 10 the productivity of resources in the long term and to quadrupling the productivity of resources in the next 20 or 30 years in the industrialised countries".

In fact, development (current human activity) is made with the environment (nature). All the natural materials and resources used in an economy end up sooner or later in the emissions or waste discharged into the environment. Every kilo of material in a finished product represents many more kilos of resources and energy extracted by man from the environment...

Gold :

More than 90% of the resources extracted are wasted in the production of food, machinery, vehicles and infrastructure industrialised countries release between % and % of the natural resources they use into the environment in the form of waste.

Three tonnes of non-renewable natural materials are needed to manufacture a catalytic converter for a car... between 8 and 14 tonnes for a personal computer:

And every year, almost 100 tonnes of non-renewable resources, plus more than 500 tonnes of fresh water, are consumed on average per person to maintain the current lifestyle of industrialised countries, i.e. 30 to 50 times more than is available in the poorest countries. Improving the efficiency with which our societies use natural resources is essential if we are to make progress towards sustainable development.

Structural reforms of our economies are needed, including the dematerialisation of our economy, which consists of 'the absolute or relative reduction in the use of natural resources per unit of function or service.

To measure dematerialisation, WWF has proposed the ecological footprint method, which compares the demand for and supply of the planet's/biosphere's regenerative resources.
The purpose of the ecological footprint is therefore to compare the capacity of the biosphere (to regenerate natural resources and assimilate waste) with human demand. In practical terms, this objective translates into a research question that can be quantified: how much of the planet's bio-productive capacity is needed to meet the demand generated by human activity? This makes it possible to visualise the impact of human activity on the planet's capacity to support life.

I . Classification of natural resources

1. **Natural biological resources** are: the water we drink, the soil we cultivate, the air we breathe, the forests that provide oxygen to the atmosphere, and all plants and animals.
2. **Natural energy resources** are, by definition, those that are exploited to produce energy. They include air, sunlight, water, geothermal energy, plants and fossil fuels such as coal, oil, natural gas and uranium. Of course, we can always reduce the properties of biological resources to energy effects: isn't the food we eat transformed into chemical energy, which our muscles in turn transform into mechanical energy? But we will maintain the distinction, arguing that the energy implications of biological resources are indirect.
3. **Renewable and non-renewable natural resources**

A distinction is often made between renewable and non-renewable resources. Renewable resources are either those that regenerate naturally, or those that are in virtually unlimited supply. It is therefore essentially the preservation of their quality that interests us here, the essential question being to know the thresholds of tolerance to pollution for each of them. It should be pointed out, however, that this classification is generally made from a global point of view, because if we think on a local scale, it is quite clear that not all territories have the same reserves (think of water, in particular). The other so-called i-renewable resources are limited in quantity, and it is therefore not only the management of their quality, but also that of their quantity, that is important to us. Assessing the reserves of renewable resources is therefore a central issue, the resolution of which depends on a number of factors. For example, in the case of oil, do we base our calculations on known deposits, or do we include deposits that have yet to be discovered, which raises the question of the financial capacity to find and exploit deposits, or the short- or medium-term economic interests that may be involved in putting the brakes on the search for new deposits, etc.? Another example: uranium is a limited resource insofar as we currently use mainly uranium-235 to produce nuclear energy, and the enrichment of natural ore produces 0.71% uranium-235 compared with 99.28% uranium-238, which is non-fissile. With our current nuclear technology, therefore, uranium is a scarce resource; if we were able to use uranium 238, or thorium, the reserves would become virtually unlimited.
The two distinctions (biological or energy resources, and renewable or non-renewable resources) cannot be superimposed: a resource can be biological and renewable (air), biological and non-renewable (Mediterranean bluefin tuna, very soon), energy and renewable (the sun), or energy and non-renewable (coal).
The conservation of natural resources is only incidentally concerned with non-renewable

resources (coal, mineral oil), which are subject to the law of limitations. When known reserves are exhausted, there are only 2 possible remedies: the discovery of other deposits or the discovery of a substitute product.

The following considerations apply primarily to renewable natural resources. Renewable Natural Resources are closely linked to ecosystems e.g. water and forests or rivers.

4. Natural resources, both point and diffuse (widespread)

This classification is based on geographical concentration, in other words, is the natural resource localized in small points or does it occupy large areas? For example, forests that cover large areas can be considered diffuse resources. Point resources are concentrated and never occupy a significant area on the geographical map, as is the case for most minerals.

It is possible to combine certain classifications based on different criteria

Example: regeneration rate and geographical concentration.

	RENEWABLE	NON-RENEWABLE
DISPLAY	- Vegetation, soil, forests, bodies - Animals - Water	- Peat - Certain short-land substances e.g. gravel, sand
PONCTUAL	- Certain crops/animals requiringconditions special	- Many minerals e.g. gold

RN according to their degree of reliability (accessibility), obstructability and legality (Ross, 2002)

According to Ross (2002) reliability or accessibility is the ease with which a NR is extracted or transported. Accessible resources include alluvial rock, agricultural products and timber. On the other hand, deep rock, oil and natural gas are less accessible.

Some RN require appropriate manpower or technology for their extraction/evacuation. With regard to the mode of transport or evacuation of RN, diamonds or drugs can be taken directly and evacuated by air without any obstacle, whereas substances requiring trains or lorries pose an obstacle. The most demanding substances, such as gas and oil, are transported by pipeline.

This classification is important for understanding the causes of conflicts in different parts of the world.

RN from an economic point of view

Natural resources play different roles in the growth of states and are therefore considered differently depending on the case. For some nations, it is only forests, agricultural products or minerals, for others it is a combination of NR.

However, it is practically difficult to estimate the exact monetary economic value of certain RN when this is indirect (e.g. air) or intrinsic (e.g. biodiversity).

L. Renewable Natural Resources

1. Water

A molecule of water is made up of two smaller hydrogen atoms and one larger oxygen atom. The formula for a molecule of water is H2O. Water can be liquid, gaseous or solid.

Water covers 79% of the earth's surface. The greatest quantity of water is found in the oceans. The oceans contain 96.5% of the earth's water volume. 1.74% exists in the form of

glaciers and eternal snow and 1.7% in groundwater. The rest of the water is found in lakes, the atmosphere, the soil, marshes and swamps, rivers and streams, and in plants.
Water is either acidic or basic. Water in which the concentration of hydrogen ions (H+) is high is acidic. Water with a high concentration of hydroxide ions (OH-) is basic (alkaline). One litre of pure water at a temperature of 23.9°C contains 0.000.000.1 (10-7) grams of hydrogen ions. The same litre of water will contain the same number of grams of hydroxide ions, i.e. 10-7 g. This water is said to be neutral. The pH scale goes from 1 to 14. A pH below 7 is acidic and a pH above 7 is basic. Very acidic water (pH<5.6) or very basic water (pH>9) is not suitable for aquatic life .[1]
Water is salty when its salinity is >2.47%. It is soft when its salinity is less than 0.15%. Between the two levels of salinity, the water is said to be brackish. In relation to salt water. Fresh water is water suitable for human consumption. But fresh water represents a small proportion of the total volume of water available. Dirty water makes up 97.446% of the total volume of water worldwide. The largest quantity of freshwater is found in the form of glaciers and eternal snow (68.7%). Groundwater contains 97.74% of liquid freshwater. 1.47% is found in lakes, 0.78% in soil and 0.01% in rivers and streams.

Water in the Democratic Republic of Congo

2/3 of the DRC's surface area (2,345,409 km^2) is drained by the Congo River and its tributaries With the exception of the north-east (the Nile basin) and the coast, the Congo river basin covers 3,822,000 km. Between Lake Tanganyika and the river mouth, the river is 4,700 km long. Its average flow is 45,000 m2 per second. The bodies of water represented by the vast network of rivers, flood plains and lakes cover some 86,080 km2, or 3.5% of the national territory. The DRC's main lakes are transboundary: Albert and Edouard (Uganda), Kivu (Rwanda), Tanganyika (Burundi and Tanzania) and Moero (Zambia). Lakes Tumba and Inongo are inland. The DRC has the largest volume of fresh water in Africa. It abounds in a resource which more and more countries need. The demand for water in the Congo is a challenge for politicians and the congolese scientist to define policies mastering data to deal with a crisis possible demand for water by other countries.
Water is a natural resource that is essential for life. It is the largest component of blood and sap; without sap and blood, most plants and animals (including humans) cannot survive. Water is the natural habitat for aquatic species (e.g. fish, amphibians and algae). Thanks to the water cycle, water plays a part in regulating the climate.

2. The floor

The soil is the soft part of the lithosphere; it is the epidermis of the emerged land (Robert, 1996). For the world as a whole, potentially arable land covers around 3,200 million hectares (32 million km^2), half of which is cultivated (Maldague, 1997). Without soil, agriculture is impossible.
A soil profile is characterised by a succession of layers known as horizons (Fig. 2). A soil has 4 horizons: A, B, C and R. Horizon A, the surface horizon, contains organic matter. It is often depleted in colloids (particles less than one micron in size) and iron by leaching. The B horizon differs from A in that it is enriched in colloids and iron. The C horizon is a more or less altered original material. The R horizon is the parent rock.

[1] The pH of Lake Kivu is 9.1-9.5, while that of Lake Tanganyika is 8.0-9.0.

Soil structure consists of clay (0 to 2 um). silt (2 to 50 um). sand (50 to 200 inn) and coarse sand (200 to 2,000 um).
Biologically speaking, soil is a living environment. Bacteria represent more than 10 million individuals per gram of soil. Fungi and actinomycetes (micro-organisms closely related to bacteria) number around 1 million. To the soil microflora panorama, we should add lichens, especially on the soil surface, and mycorrhizal fungi that live in symbiosis with the roots. The pedofauna includes earthworms, mites and burrowing animals (rodents, insectivores, etc.). Living organisms play an essential role in both soil formation and soil functioning. For example, bacteria help decompose plant debris to form humus. Living organisms also participate in the degradation of rocks and minerals, thereby contributing to the formation of soil.
Five factors influence soil formation. These are climate, vegetation and living organisms, bedrock, relief and soil age.
Climate. This is the major element in soil formation, because in every climatic zone of the world, soils have common characteristics (Clibbon, 1996).
Vegetation and living organisms. The decomposition of plant and animal debris (humification) results from the activity of micro-organisms. This transformation results in the formation of humus.
Bedrock. This is a loose deposit that gives rise to soil. It is different from bedrock
Relief. This has a profound influence on soil development. There is a difference between soils on summits, slopes and lower slopes. The phenomenon of leaching generally predominates on higher ground, while the phenomenon of accumulation is more prevalent in depressions.
The age of the soil. The influence of climate, vegetation, bedrock and relief depends on how long these factors have been in action. Soil takes a long time to form (several hundred years). Soil is therefore a resource that is difficult to renew. Maldague (1997) notes that soil degradation is often irreversible. Robert (1996) considers soil to be a non-renewable natural resource.

3. Vegetation

Forests, wooded meadows and shrublands cover 53 million km^2 , or about 40% of the Earth's land area (IUCN et al., 1991). According to an estimate by Grandtner (quoted by Clibbon, 1996) there are between 50,000 and 60,000 species of trees on Earth.
Around 240 million of the world's poor live in the forests of developing countries and depend on forest products for their survival. The forest and its products provide cash income, jobs, and consumer items for poor families. Forest exploitation provides formal and informal employment for around 40-60 million people. In some developing countries, the sector contributes more than eight per cent of GDP. Timber may be the most important forest product, but forests are also frequented for fruit, herbs and honey, as well as for wild animals.
Less visible but no less important are the ecosystem services provided by forests, in this case the hydrological cycle.
Natural and modified forests provide numerous benefits to human beings. They are an integral part of life-support systems and play a crucial role in balancing the atmosphere and climate. Forests play a moderating role in the local climate. Climatic conditions are

generally milder, wetter and more stable in areas under the influence of forests than in non-forested areas of the same region. Forests regulate the local water cycle; they protect soils from excessive erosion, reduce the sediment load of watercourses, slow run-off and moderate floods and other common fluctuations in water flow. In catchment areas, the presence of forests also has a regulating influence on run-off and helps to conserve spawning and fishing habitats.

Forests provide a habitat for a wide range of species. In the forest of the Kahuzi-Biega National Park in the DRC, there are 13 species of primate, 9 species of antelope, more than 400 species of bird and several thousand species of plant (Muhlenberg, 1995).

Forest genetic resources are used to improve species grown for timber; they include wild species closely related to animal breeds and plant varieties selected for breeding and horticulture, as well as species that can offer new products.

Forests play an important role in tourism. They have an inestimable cultural value as settings for leisure and relaxation and as a source of religious and artistic inspiration.

Nevertheless, global forest cover has declined by at least 20% since pre-agricultural times. While the forest area has increased slightly over the last thirty years in industrialised countries, it has declined by around 10% in developing countries over the same period.

According to the FAO, deforestation causes 25% of greenhouse gas emissions, and reducing this is a high priority on the global agenda.

4. Wildlife

Wildlife, as a resource, can be linked to animals domesticated by man for various purposes: food, source of mechanical energy, protection, companion of man, spiritual value and source of various products. Wildlife plays an important role in human consumption. 80% of the meat consumed in West Africa comes from the slaughter of wild animals.

Other roles played by wildlife are :

- Pedofauna and burrowing organisms dig galleries that aerify the soil. Their dejections fertilise the soil.
- Certain insects are essential for pollination and for biological control of crop diseases. In the USA, 99% of insect pests are controlled by their natural enemies (Pimentel et al., 1997). Each insect pest has between 10 and 15 insects against it.
- In the pares, animals are a tourist attraction.
- Through their interaction with plants, animals play an important role in the germination of certain species. In Kahuzi-Biega National Park, Myrianthus holstii (one of the main plants eaten by gorillas) germinates after the seeds have been ingested by gorillas or elephants.
- Some specimens are used by zoos for biomedical experiments.
- Animals are used in the circus.

In conclusion, all natural resources are very important. Water, soil, vegetation and fauna do not exist in isolation from each other. There is a complementarity between these different types of natural resource. The forest plays a part in the water cycle and climate regulation. The climate favours soil formation. Animals contribute to soil formation and forest regeneration. The soil is the substrate for plants. The different types of natural resource must therefore be considered in relation to each other. These are the components of ecosystems. Ecosystem conservation is therefore an excellent way of conserving natural resources.

M. Ecosystems

The conservation of natural resources requires that the biotope and associated biocenosis be considered as a functional whole (system). The ecosystem or natural environment is a structural system made up of a biocenosis (a collection of plants, animals and other organisms) and a biotope (non-living elements of the environment).

1. Types of ecosystem

IUCN (The World Conservation Union) et al (1991) define 5 types of ecosystem in relation to human impact on these environments

- **The built ecosystem**: where buildings, roads, airports, docks, dams and other human infrastructures predominate. This category also includes parks, gardens, golf courses and other urban and suburban green spaces.
- **Cultivated ecosystem**: an ecosystem in which the human impact is greater than that of any other species, and most of which is cultivated.
- **Degraded ecosystem**: an ecosystem whose diversity and productivity have been reduced to such an extent that its regeneration requires active rehabilitation or restoration efforts.
- **Modified ecosystem**: an ecosystem in which the human impact is greater than that of any other species, but most of which is not cultivated.
- **The natural ecosystem: an** ecosystem where, since the industrial revolution (around 1750), human impact (a) has not been greater than that of other indigenous species, and (b) has not affected the structure of the ecosystem.

2. Types of natural ecosystems

There are a variety of environments, inhabited by plants and animals. There are terrestrial and aquatic ecosystems.

- **Terrestrial ecosystems** include forests (tropical rainforests, boreal forests, gallery forests, bamboo forests, etc.), savannahs, deserts, etc.
- **Aquatic ecosystems** include rivers, lakes, seas and wetlands (marshes, swamps, mangroves, peat bogs, etc.). Water is the predominant natural resource in an aquatic ecosystem. Each type of ecosystem has its own characteristic species. In the mountain rainforests of the Kahuzi-Biega National Park (between 900 and 2,300 m altitude), a terrestrial ecosystem, we find the plant species Albizzia gummifera, Parinari excelsum, chrystophyllum gorugosanum and the animal species Gorilla gorilla graueri, Loxodonta africana cyclotis. Lake Kivu, an aquatic ecosystem, is home to 26 species of fish (Basina, 2004).

The intermediate zone between two ecosystems is called an ecotone. Ecotones are usually richer in individuals and species than neighbouring biocenoses. For example, shorebirds are specifically more diverse than those of the open sea and adjacent land.

N. Functioning of natural ecosystems

Food relationships play an important role in the functioning of ecosystems. The concept of the food chain must be mastered to understand how ecosystems function.

The food chain represents the list of plant or animal species in the relationship "such and

such eats such and such" or "such and such is eaten by such and such" (Muhigwa, 2001).
There are 3 types of food chain: the predator chain (or epigeic chain), the parasite chain and the decomposer chain.
The chain of predators begins with chlorophyllous plants, which are the producers. Then come the first-order consumers, the herbivores, who eat the producers. These first-order consumers are themselves eaten by second-order consumers, the carnivores. This is a typical chain in which the successive components are progressively larger in size but smaller in number.
Example: In a lake, numerous species of algae that thrive on solar energy and nutrient salts in solution in the water serve as food for different species of planktonic crustaceans, which in turn serve as prey for zooplankton-eating fish, which in turn are eaten by carnivorous fish, which are also eaten by humans.
The chain of parasites goes from large animals to smaller and smaller animals. For example, vultures that feed on antelope meat left behind by a lion.
The chain of decomposers originates in dead organic matter from plants, animals and micro-organisms. The decomposers are known as bio-reducers. This dead organic matter goes through phases of humification and mineralisation by the effect of bio-reducers (bacteria, fungi). The endogeous chain plays a part in the primary productivity of the environment, as it provides producers with the mineral elements that are essential for their growth.
N.B. An animal can belong to several different trophic levels. This is the case with omnivores, which eat both animals and plants.
In these predator-predator successions, a change in one of the links in the ecosystem can lead to a reduction in the population of planktivorous fish that prey on the predators. The introduction, for example, of a carnivorous fish into an area that is devoid of carnivorous fish can lead to a reduction in the stock of planktonivorous fish that prey on predators. Planktonic crustaceans will then be subject to significant predation by fish. The increase in their numbers will increase predation, which will reduce stocks. The reverse is also true: when, for example, a carnivore is overexploited and becomes rare in a lake where it was previously abundant, the predators will increase in number and reduce the zooplankton, which obviously leads to an abundance of phytoplankton.
In this way, it has been possible to demonstrate the existence of key species, which play a more important role than others and whose disappearance would have multiple consequences (Leveque, 1994). This is the case for certain species that play a decisive role in plant pollination, herbivores and predators whose presence ensures the coexistence of a large number of species by controlling the development of invasive species.

12 . Economic value of ecosystems

Ecosystems provide enormous services to humanity. The monetary value of ecosystem services worldwide has been estimated at an average of 33,000 billion US dollars (Costanza et al., 1997). For the United States, the value of the services provided free of charge each year by ecosystems is 319 billion US dollars (Pimentel et al., 1997).
The services provided by ecosystems include climate regulation, gan regulation, disturbance regulation, water regulation, water supply, erosion control, soil formation, nitrogen fixation, waste treatment, pollination, biological control, habitat, food

production, raw materials, genetic resources, recreation (tourism) and culture.

The logic behind these estimates is based on government intervention in the absence of the ecosystem service. If the ecosystem providing a service no longer existed, the state would invest money in providing this service. The fact that this service is provided free of charge means that it is less of an expense for the government.

Ecosystem services are very expensive. American researchers have tried to create a 3 km artificial biosphere in Arizona in the United States. This biosphere worked, but its services were three times more expensive than those of our natural biosphere.

It is true that the conservation of natural resources and ecosystems is of vital importance to our development. Unfortunately, this importance does not protect them from destruction.

O. Ecosystem threats

The main threats to natural resources and ecosystems are linked to natural and man-made factors.

Natural factors include climate change, wildfires (especially in boreal environments), ocean water level fluctuations, volcanic eruptions, meteorite impacts and disease.

Examples of species extinctions due to natural phenomena (Lavoie, 1998)

At the end of the Cambrian (500 million years ago), a species extinction occurred. It is thought to have been caused by the displacement of the Gondwana super-continent (plate tectonics) towards the South Pole, triggering a prolonged glaciation during which species were extinguished. The worst extinction of species occurred in the Permian (250 million years ago). It is thought to have been caused by volcanic eruptions that threw more than 2 million km of lava onto the surface of the planet, creating a gigantic dust cloud with catastrophic consequences. In the Cretaceous period (65 million years ago), the collision of meteorites (large-scale extraterrestrial objects) and the Earth also led to the extinction of several species.

At the height of the ice ages, forest areas were significantly reduced, with only certain more favoured regions remaining, known as refuge zones. These refuge zones are located in tropical environments, especially in South America and Africa.

The threats posed by natural factors have been catastrophic. Today, man-made threats are causing huge losses to ecosystems and natural resources. Tropical deforestation, the destruction of wildlife, soil degradation and water pollution are just some of the causes.

P .*Some threats to natural resources from man-made factors* :

1. Tropical deforestation

When it comes to recent losses of biodiversity on a global scale, the phenomenon of tropical deforestation is singled out as being responsible for these losses (Lavoie, 1998). Tropical forests are being destroyed on a massive scale and at a rapid pace. Tropical rainforests, which are home to more than 50% of all living species, have declined considerably over the 20th century. Deforestation is estimated at 11 million hectares per year. The clear tropical forest is disappearing.

Logging, slash-and-burn agriculture and the search for fuelwood are the main causes of tropical deforestation. If, using current methods, 2% of the woody stems with a diameter at breast height of more than 10 cm are extracted from a tropical forest (test plot measuring 300 m x 300 m), 26% of the trees of the same category are damaged, the forest cover is reduced by more than 50% and machinery affects 8% of the surface area of the forest.

Q. The consequences of deforestation are the transformation of forests into savannah, soil erosion, the disappearance of species and regional climate change:

1. Transforming a forest into a savannah

When abandoned, it is difficult to return plots of tropical forest to their original state, even in the long term. For example, the 3.5 million hectares of tropical forest in the eastern Amazon that were transformed into farmland and abandoned in the first half of the 20th century are now, more than fifty years later, revegetated with grass and shrub species. Soil erosion, the loss of nutrients through burning and intensive agriculture, and the removal of tree seedbeds are partly responsible for this situation (Buschbacher, 1986; Daily, 1995, cited by Lavoie, 1998).

2. Soil erosion

Exposed tropical soils degrade rapidly and irreversibly, through ferralitisation, oxidation and increased erosion. Eroded soils have a reduced capacity to retain water, leading to more severe flooding in regions bordering rivers that draw their water from other deforested regions (William, 1990: cited by Lavoie, 1998).

3. Disappearance of species

It has been calculated that with an annual deforestation rate of 1% in the world's tropical forests, between 0.2 and 0.3% of the species in these forests are lost each year. If we conservatively estimate that there are around 2 million species in tropical forests, this means the loss of between 4,000 and 6,000 species. The massive disappearance of species in tropical environments can have repercussions in temperate and boreal environments. For example, the spruce budworm epidemics that affect eastern Canada are controlled in part by this species' predators, the insectivorous birds that are present during the summer season. These birds migrate in autumn to tropical regions (mainly Central America and parts of South America). It has been estimated that the disappearance of 75% of these birds (in terms of numbers) would result in a marked reduction in their influence on budworm populations. Radar images capturing the flights of migratory birds over the Gulf of Mexico indicate that the frequency of these flights has recently fallen by 50%. If the trend continues, bird populations could, in the near future, be at a precarious level preventing them from controlling budworm populations, with serious consequences for the Canadian forestry industry (Shuka et al. 1996; quoted by Lavoie, 1998:144).

4. Regional climate change

The models generally predict an increase in mean annual temperature of between 0.3 and 3.8°C, a fall in total annual precipitation of between 220 and 640 mm and a decrease in evapotranspiration by plants (between 164 and 500 mm of water on an annual basis). In most simulations, the fall in precipitation is not offset by the fall in evapotranspiration, which could result in longer dry seasons. Such dry seasons would make it difficult for the forest to recover, even after abandonment (Shuka et al. 1990 quoted by Lavoie, 1998).

R. The destruction of wildlife

American bison, which numbered an estimated 75 million before the arrival of the first

Europeans, were exterminated in the 19th century; only a few individuals remain in parks (Leveque, 1994). In Africa, in Benin, the number of individuals of some animal species fell between 1978 and 1989 as follows: buffalo, from 5,815 to 2000; hartebeest, from 4,030 to 1,300 and waterbucks, from 1970 to 700 (Ofaumon, 1995). In the DRC, in the high altitude part of the Kahuzi-Biega National Park (600 km^2), the number of gorillas has fallen by 50% and the 771 elephants counted in 1996 almost disappeared in 2000 (Inogwabini et al. 1996; WCS, 2000). In the case of these two countries, this is due to pressure from poaching.

The fauna of the Aral Sea in Kazakhstan has almost completely disappeared, in particular almost all of the 24 endemic fish species (Leveque, 1994). This disappearance is linked to pollution of the lake.

The Nile perch Lates niloticus ("Zukolye"), one of Africa's largest predatory fish, introduced into Lake Victoria in the 1950s, is thought to be one of the main causes, along with over-intensive fishing, of the destruction of a rich endemic fauna of several hundred species of small fish belonging to the cichlid family (Leveque, 1994).

The main causes of wildlife loss are habitat destruction, inappropriate hunting techniques, poaching, pollution of aquatic environments, inadequate fishing techniques and the introduction of exotic species.

The loss of wildlife species causes an ecological imbalance in the ecosystem, with harmful consequences for the economy. Disturbances in the food chain are likely to result in the death of predators of the extinct species. Symbiotic relationships between animals and plants can break down, causing problems for the reproduction of certain species. Certain animals play a vital role in the regeneration of plants and the consumption of invasive species. Their disappearance would lead to the proliferation of these invasive species and regeneration would diminish. The genetic potential of wildlife can be used to improve domestic animals. Many medical trials are first carried out on animals before being extended to humans. In economic terms, the destruction of wild animals leads to a loss of income from their exploitation and food.

Soil loss is one of the major problems affecting the biosphere. Globally, it is estimated that 0.7% of the planet's soil capital is lost each year (Leveque, 1994). According to Maldague (1997), it was predicted that by the end of the 20th century the area cultivated per capita would have fallen by 50%. Some 600 million hectares of arable land will have been damaged by soil erosion, salinisation, waterlogging and urban expansion. Of the more than 250 million hectares of irrigated land in the world, 10 million are abandoned each year, and half of these are condemned to secondary salinisation.

a. Erosion

It is defined as the action of removing by gnawing or the result of this action. By extension, erosion is any removal of rock or soil material by external agents: water, wind, frost, glaciers, rivers, the sea, etc. In tropical environments, a distinction is generally made between water erosion (rain erosion) and wind erosion (wind erosion). In desert environments, wind erosion predominates. In our environments, tropical forest and savannah, erosion by water is common. There are 4 stages in the process of water erosion (Maldague, 1997).

- ***Falling*** water (ramming and damaging of soil by water).

This phase can be described as "splash erosion". This phase is characterised by the splashing of small soil particles into the air. This projection leads to the disaggregation of the lumps that make up the soil and are responsible for its structure. The destruction of the aggregates will be all the more rapid and intense if the soil has a poorer structure, i.e. if it has fewer binders [the importance of organic matter] between the elementary particles.

- ***Sheet erosion***

Sheet erosion consists of the removal of soil particles by water flowing over the entire surface of the soil. In the case of denuded soil, when it rains, the drops, after having hit the ground with more or less considerable force, begin by infiltrating, but this infiltration is quickly followed by a run-off phase which causes the soil particles to move. If there is the slightest slope, the current creates a pressure that breaks up the soil aggregates; the constituent particles are dragged along and transported by the run-off water to the watercourses, where the water becomes turbid and takes on a yellow or brown hue depending on the nature of the particles in suspension. Turbidity is the result of physical pollution of the water by solid matter, which reduces the level of dissolved oxygen. Sheet erosion is due to generalized runoff over the entire surface of the soil.

- ***Erosion in gullies***

Under the effect of the slope, the drops of water form nets, then small rivulets that run down the slope, digging into the soil and making small furrows: At this stage, grooves, gullies and ravels are formed, causing the top layer of soil to be stripped away on the slopes. Fine elements, ash, soluble salts and organic debris are carried away. The result is intensive leaching.

- ***Digging erosion***

At the bottom of the slopes, where the muddy water has acquired sufficient energy, it opens up wide crevasses in the ground: this is the digging erosion that is characterised by the formation of ravines.

In the case of a watercourse, bank erosion is more intense on the concave side - where the speed of the water is greater - than on the convex side. In this case, each time there is a flood, clods of earth, sometimes even whole sections of the bank or boulders, are scoured and fall into the river bed.

The main factors influencing sheet erosion are (Clibbon, 1997) :

- the gradient of the slope, i.e. its inclination to the horizontal, and its length;
- the nature of the tographic surface and in particular its roughness (presence of obstacles to runoff)
- the distribution over time of rain showers, which contribute to more concentrated runoff; also, the intensity and duration of rainfall;
- the type of groundwater runoff (laminar or turbulent flow);
- the rate of infiltration of water into the soil. This depends above all on the physical characteristics of the soil, and in particular the thickness and permeability of the A horizon;
- repeated fires, which consume humus ;
- current land use and, in particular, the nature, density and continuity of plant cover during periods of rainfall.

The probability of erosion becomes very high when at least two of the three risk factors are present: deforestation, unstable structure and steep slope. The biggest cause of erosion is deforestation: soil that is not protected by plant cover is quickly eroded. Experiments carried out in French Guiana between 1979 and 1987 show that :

- Under a forest, soil stability is high and erosion is low (50 to 70 kg/ha/year);
- mechanical clearing multiplies soil transport through erosion (Table 1).

Effect of deforestation on soil erosion

Deforestation method	Runoff (in mm)	Erosion at ground tonnes/ha
Control drill	0,0	0,00
Traditional	6,6	0,02
Manual	47,7	5,00
Clear cut	104,8	4,80
Mechanised (for pushing and pulling up trees)	250,3	20,00

Source: Lal (1990, cited by Robert, 1996: 91)

Other causes of erosion are :

- **Compaction**

Agricultural mechanisation causes the pores in the soil to disappear. This makes the soil impermeable, encouraging surface run-off.

- **Overpaturation**

When the carrying capacity of a plot of land is exceeded due to a very high density of domestic or wild animals, there is excessive grazing and trampling. This leads to loss of plant cover and soil compaction.

- **Inadequate farming techniques**

Ploughing in the direction of the slope and burning, for example.

b. Soil salinisation

A soil is said to be dirty when its salt concentration exceeds 500 mg/l (Robert, 1996). A distinction is made between natural salinisation (primary) and anthropogenic salinisation (secondary), generally linked to irrigation,

Primary salinisation is linked to a specific natural source of salts, which may be the sea or the ocean, but also saline rocks and sometimes the alteration of rocks and volcanism.

From the source, salts such as chlorides (Na, K, Ca, Mg) or sulphates (Na, Ca, Mg), which are highly or moderately soluble respectively, can move and contaminate both groundwater and soil.

Secondary salinisation is linked to human activity, mainly irrigation. Irrigation is a common practice in arid and semi-arid regions where water is the limiting factor for crops. The causes of secondary salinisation correspond to simple geochemical mechanisms. In fact, even if irrigation is initially carried out using water with a low salt content (< 500 mg/l), the high levels of evaporation and above all evapotranspiration from crops in arid areas lead to salt concentrations in the surface zones of the soil. If we irrigate with quantities of water that significantly exceed the evapotranspiration values, the risks are salinisation and the rising of the water table in the soil profile.

c. Urbanisation

Urbanisation covers the ground (of cities) with roads, railways, airfields, industrial installations and more or less impermeable pavement. Precipitation water is collected by sewers and conveyed directly to rivers or the sea. It is estimated that more than one million km^2 of the world's urban area is thus deprived of rainwater infiltration (Basabose, 1999). This loss of water causes the soil to dry out.

The consequences of soil degradation can be catastrophic for mankind. Economically, as soil is the main natural resource for agriculture, the reduction in arable land is leading to food shortages for an ever-growing human population. Dirty soils are unfit for cultivation. Salts exert a high level of stress on plants, which may be linked to high osmotic pressure and therefore reduced water availability (corresponding to reduced water activity) or to the direct toxicity of salts (physiological effects of sodium on plants). Salinity destabilises soils by encouraging the dispersion of clays.

There is evidence that salinisation phenomena linked to irrigation may have played a role in the geographical displacement or decline of prehistoric Mesopotamian and Egyptian civilisations (Robert, 1996).

Erosion can sometimes considerably reduce the capacity of reservoirs, making the construction of dams pointless. Sedimentation due to soil erosion is the deadly disease of dams >> (Maldague, 1997).

In ecological terms, soil degradation means the destruction of the organisms and species that live in it or use it as a substrate. As a result of soil erosion, water is polluted, threatening aquatic fauna and flora.

S. Water pollution

Water is polluted when it contains impurities that impair the use of the resource. Impurities in water are represented by suspended matter and dissolved matter. There are two types of water pollution: diffuse pollution and point source pollution.

a. Non-point source pollution

Diffuse pollution is a group of pollutions whose source is not easily identifiable or perennial, but whose sum can have disastrous consequences (Rodriguez, 1997). Diffuse pollution is generally linked to agricultural activities. But it is also found in cities. It affects surface water (rivers, lakes and seas) and groundwater.

V.The anthropogenic factors that accelerate diffuse pollution are :

- agricultural runoff, which carries nitrates and phosphates;
- urban run-off, which carries nitrates, phosphates and inorganic elements; - landfill sites, from where substances percolate into the water table;
- erosion, which leads to the sedimentation of water bodies and the input of nutrients; - atmospheric pollution by minerals;
- recreational activities, by hydrocarbons and fumes.

Natural fertilisers (manure, slurry, compost), chemical fertilisers (N-P-K) and pesticides are the main sources of diffuse pollution.

Surface water pollution

- Phosphates are a problem in surface water. Phosphorus in water, especially stagnant lake water, encourages the proliferation of algae and a large development of bacteria whose activity reduces oxygen, resulting in a shortage of oxygen in the lake. This phenomenon is known as eutrophication >>.

Eutrophication is a natural phenomenon in which aquatic ecosystems age. Normally, freshwater ecosystems are nutrient-poor at the beginning of their existence. Over tens of thousands of years of accumulating nutrients from the surrounding land, they become progressively richer.

As a result of human activity, the discharge of nitrogen and phosphate fertilisers is rapidly eutrophying ecosystems.

Eutrophy is the almost terminal stage in the life of a lake. Eutrophic lakes are often shallow lakes. The water in eutrophic lakes is very rich in nutrients. Phytoplankton is very abundant in the surface layer. Algae overgrowths (clouds) are frequent and littoral flora is abundant and dense. The bottom is made up of putrescible substances. There is often a prolonged dissolved oxygen deficit in the lower layers of the hypolimnion. The fauna of the deep layers is poor in species, as fish cannot frequent this low-oxygen zone. In these anorexic conditions, species that need oxygen to live, especially fish, die of asphyxiation. This is how more than 266 species of fish have disappeared in Malaysia (Leveque, 1994).

Groundwater pollution

Nitrates are the main cause of groundwater pollution. When they are applied to fields in permeable soils, the nitrates not used by the crops percolate down through the pores into the aquifer. Excess nitrate in water can cause poisoning in humans and animals. Nitrate poisoning leads to abortion of the pores. When groundwater is polluted, it has to be abandoned because no one has yet been able to clean it up (Rodriguez, 1997).

The biggest problems of diffuse pollution come from pesticides. The DDT (dichloro-diphenyl-trichlorethane) used in agriculture accumulates in organisms throughout the trophic chain, altering their physiology and endangering their survival because of its toxic effects. This is how beluga whales, cetaceans from the Gulf of St Lawrence in Canada, were rendered sterile for several years.

Point source pollution

Diffuse pollution is generally linked to agricultural activities and the source is diffuse, making it difficult to detect. Point source pollution results from the direct discharge of a contaminant into the water. The source can be identified. Point source pollution is mainly urban.

The sources of point source pollution are effluents from factories, industries and residences. Pollution comes from sewers.

Diffuse pollution brings inorganic substances into the water, such as fine elements (aluminium, zinc, copper, etc.), which are dangerous for aquatic life and drinking water. Also, liquid or solid waste carried by sewage is loaded with pathogenic micro-organisms (bacteria, for example) that affect recreational uses.

There is a great deal of oil pollution in the marine environment. It is estimated that 600,000 tonnes of hydrocarbons of various origins are discharged into the oceans every year (Leveque, 1994). These accidents, which cause oil slicks, lead to the death of animals.

Case study: the Lake Kivu coastline

Basima (2001) carried out a physical and biotic characterisation of the shoreline of Lake Kivu, Bukavu basin, western zone. The site studied was located between the Bwindi River and Ndendere Bay. Results. The Bwindi, Chula and Wesha rivers often carry mud. The Kawa river carries a lot of urban waste and sewage. During heavy rains, as a result of the quantities of mud brought in, the water is brown in colour up to 300 m offshore and 100 m along the coast to the west. 20 of the sites studied lie between the fields from which the soil sedimenting the coast comes. Seven steeply sloping sites are completely bare. The banks are affected by slumping. There is more silt on the steep slopes than on the gentle slopes. There is a negative correlation between the presence of fish and silt. There is a positive correlation between silt and erosion. Silt covers spawning sites for fish and disturbs territories for coastal fish.

W . Threats to natural resources from the global effects of environmental degradation

The global effects of environmental degradation are the greenhouse effect, the thinning of the ozone layer and acid rain.

1 The greenhouse effect

The atmosphere plays the same role in relation to the earth as a glass pane in a greenhouse (Leveque, 1994): light passes through the pane and heats the inside of the greenhouse, which in turn emits infrared radiation. As the infrared radiation cannot penetrate the glass, the heat remains trapped and the temperature inside the greenhouse rises. By the same token, the ground that receives the sun's rays emits infrared radiation in return, around 40% of which is intercepted by the so-called greenhouse gases present in trace form in the atmosphere, while the rest is dissipated into space. These gases, such as carbon dioxide, methane and CFCs, play a vital role in the current climate balance, as they contribute to the significant warming of our planet's surface, making life possible. Without them, the average global temperature would be -18°C instead of the current +15°C. But increasing them too much can also have the opposite effect. The increase in greenhouse gases in the atmosphere is already responsible for an estimated global warming of 0.5°C since 1880. If the increase in greenhouse gas concentration continues, the average global temperature could rise by 4 to 5°C in the space of a century.

Warming seawater could be the main factor behind the coral bleaching that has been observed around the world since the early 1980s. This phenomenon, which is often fatal, corresponds to the loss of coloured organisms, the zooxanthellae, which live in symbiosis with the coral. Warming due to the greenhouse effect could change the locations of major ecosystems, as happened during the last glacial-warming cycle.

The use of fossil fuels (coal, natural gas, oil) and the destruction of forests, in particular large areas of tropical forests, are the main causes of the increase in CO and its accumulation in the atmosphere. In 1988, 5.5 billion tonnes of CO_2 were added to the atmosphere, and between 0.4 and 2.5 billion tonnes more were added as a result of deforestation (Maldague, 1997).

2 . The thinning of the ozone layer (Villeneuve, 1998)

Ozone is a particularly unstable molecular form of oxygen (formula: O). It is a gas found in the stratosphere, a layer of the atmosphere 35 to 55 km above the ground. Ozone acts as a filter for ultraviolet radiation in the upper atmosphere. It absorbs ultraviolet (UV) radiation, which has enabled organisms to colonise the earth's environment. Since 1969, the O layer that used to protect the Earth has thinned by around 4%. Worse still, since 1985 it has been noted that a hole develops in the O layer over the Antarctic continent in spring (April-May). This hole is larger than the USA and deeper than Mount Everest (8.8 km), and corresponds to the loss of almost 50% of the O present in the stratosphere above the continent.

The depletion of the ozone layer results from the reaction of ozone with chlorofluorocarbons (CFCs). CFCs are released by factories, cars, incinerators and refrigeration equipment. They do not decompose in the troposphere and migrate slowly to the stratosphere. In the stratosphere, they are dispersed and decompose under the action of sunlight into individual chlorine molecules. Chlorine reacts with ozone and destroys it (o) + cl Oz+ clo). The penetration of ultraviolet rays into the atmosphere has harmful consequences for humans, plants and animals.

It is predicted that each 1% reduction in the thickness of the ozone layer would mean a 4 to 6% increase in cases of skin cancer (600,000 cases are recorded each year in the USA) and a death toll of 100 to 500 among white Americans. Unlike skin cancer, which only affects whites, the weakening of the immune system can affect all races. There is also talk of a huge increase in cataracts, the opacity of the cornea that can lead to blindness. A large proportion of domestic plants used in agriculture are sensitive to ultraviolet rays. A 25% reduction in the ozone layer would cause a 25% reduction in soya yields.

Flora and fauna are also affected. Experiments carried out on a type of microscopic algae that forms part of the phytoplankton show that its production would fall by 35% in the event of a 25% reduction in the ozone layer. This reduction would be catastrophic for aquatic ecosystems, since phytoplankton form the basis of marine food chains. The increase in ultraviolet light could significantly threaten the development of many species of marine fish and crustaceans.

3 Acid rain

As they are transported by air, sulphur (SO2) and nitrogen (NOx) dioxides and volatile hydrocarbons (from the combustion of fossil fuels) form sulphuric acid, nitric acid, ammonium salts and ozone. These substances fall back, sometimes at great distances from their point of emission, in the form of dry particles or dissolve in rain, snow, fog and reed, forming acid precipitation (Maldague, 1997).

Acid precipitation and the sudden melting of acid snow have had a major impact on the fauna of lakes and rivers in North America and northern Europe (Clibbon. 1997). It has been discovered empirically that when the pH of a lake falls to 5.5, most of the ichthyological fauna (e.g. fish) is generally in danger, mainly because of the problems this causes for reproduction. When the pH is below 5, most fish cannot survive (Stokes, 1988; quoted by Clibbon, 1997). Below pH 4.5 all fish species disappear; in general, waters with a pH >9 are not conducive to fish development either (Vaillancourt, 1968; quoted by Clibbon, 1997). At the end of 1985, approximately 7 million hectares (70,000 km^3) of the forests of 15 European countries were affected to varying degrees by acid rain (Maldague, 1997). Acid deposition affects forests either directly, by affecting the foliage, or indirectly, by altering the properties of the soil from which the trees draw the elements they need to grow. In Europe, more than 5 million hectares are threatened by acid rain.

Acid rain causes damage to property and infrastructure because acid water is corrosive and erosive.

Effects of acid rain

The Adirondack Mountains region of upstate New York is densely wooded and dotted with lakes. It lies in the path of winds from the heavily industrialized Ohio River Valley. It is a veritable source of acid precipitation. In the 1930s, the average pH of Adirondack lakes above 600 m was 6.5, with only 4% of lakes having a pH of 5. By 1975, 51% of lakes had a pH <5. During this period of about 40 years, entire populations of brook trout, lake trout, white suckers, brown bullheads and other species were exterminated (Clibbon, 1997).

- As a result of acid rain from Germany and the United Kingdom, more than 6,500 lakes in Sweden and Norway no longer contain fish, and 200 lakes and rivers in the province of Ontario (Canada) have died (Clibbon, 1997).

The current threats to natural resources and ecosystems are generally the result of human activity. It is up to mankind to limit these threats through conservation measures.

W. Classification of protected areas

There are ten categories of protected areas and biosphere reserves (Maldague, 1997; Lavoie 1998).

Category I: strict ecological reserves

Territory containing remarkable, fragile or unique ecosystems in which human intervention is prohibited (including tourism) with the aim of maintaining the various natural processes that govern these ecosystems in their entirety. Scientific research is one of the main reasons for creating this type of reserve.

Category II: National parks

Generally very extensive territories where the ecosystems found there are remarkable and undisturbed by human activity (at least, following the creation of the park). Tourism and light recreational activities are permitted. The educational function is generally quite important. Certain areas of national parks may be subject to strict protection in order to conserve highly sensitive habitats in their entirety.

Category III: national monuments

Territories protecting a very special natural feature. Such areas are often small and attract a lot of tourism.

Category IV: wildlife refuges

Areas of great local, regional, national or global importance for wildlife conservation. In general, habitats within wildlife refuges are subject to certain adjustments to favour the target species.

Category V: protected landscapes

This category contains a wide variety of natural landscapes which have been more or less altered by man and which are conserved to varying degrees for aesthetic purposes (promotion of tourism) and to serve as evidence of the harmonious use of natural resources.

Category VI : Park reserves

Territories, generally large in area and more or less inhabited, for which an official designation has not yet been granted or is not yet in force. These are often areas with a presumed high ecological value, but for which there is insufficient biophysical data to justify this.

Category VII : Anthropological reserves

Territories created to protect the traditional way of life of particular human societies. In most cases, these are indigenous groups whose small numbers mean that they have little influence on their environment. The creation of this type of reserve is therefore generally beneficial to the local flora and fauna.

Category VIII: multiple-use landscaped areas

Territories where the harvesting of natural resources (timber, wildlife) is permitted but where certain restrictions are imposed in order to manage the resources sustainably.

Category IX: Biosphere reserves

Biosphere reserves are designated throughout the world with the aim of developing a network of areas representative of the world's natural diversity, where human activities are compatible with nature conservation and sustainable development objectives. Biosphere reserves are internationally recognised as part of UNESCO's Man and Biosphere (MAB) programme. They are designed to answer the question: "How to reconcile the conservation of biological diversity, the quest for economic and social development and the maintenance of associated cultural values" (UNESCO, 1996). In practical terms, a biosphere reserve should contain three elements.

"One or more core areas with long-term protection for the conservation of biological diversity, the monitoring of the least disturbed ecosystems, and research and other low-disturbance activities (e.g. education), a well-identified buffer zone normally surrounding or adjoining the core areas, used for cooperative activities compatible with ecologically sustainable practices, including environmental education, recreation, ecotourism and applied and basic research, and a flexible transition zone (or cooperation area), which may include a number of agricultural activities, human settlements or other operations, and in which local communities, management agencies, scientists, non-governmental organisations, cultural groups, economic interests and other partners work together to manage and develop the resources of the area in a sustainable way".

The essential element of biosphere reserves is the link between biodiversity conservation and the development needs of local communities. This was recognised in Caracas, Venezuela, in February 1992 by protected area planners and managers from around the world as a key factor in the management of most national parks, nature reserves and other protected areas.

Category X: Natural World Heritage Sites

Such sites are designed to highlight the exceptional and universal value of the natural features they contain.

Number and size of parks worldwide

In 1989, the total surface area of protected territories was 4,850,000 km^2 , representing 3.2% of the earth's surface. However, only 2% of the world's land surface is preserved within strict ecological reserves or national parks. What's more, if we exclude the huge Greenland park, which alone covers 700,000 km2, barely 1.6% of the world's land surface is actually protected (Primack 1993, cited by Lavoie, 1998). According to Caughley and Gunn (1996: cited by Lavoie, 1998), 8,500 areas of more than 1,000 ha, covering 5.2% of the earth's surface, are protected.

Criteria for creating reserves

Precarity: A vulnerable, threatened or endangered species should be given higher conservation priority than a species whose populations are healthy. Similarly, a biological community facing imminent danger of destruction should be given priority for protection.

Utility: biological species or communities of appreciable value to the surrounding human populations must be protected before those of only relative value.

Criteria for creating a network of reserves

Complementarity: any reserve added to an existing network must be complementary to the other reserves in the network. Complementarity is closely associated with the notion of efficiency (E), which can be expressed by the formula $E = 1 - (x/t)$ where "x" is the number of sites or the area needed to achieve a specific conservation goal and "t" is the total number of potential sites or the total potential area where conservation measures could be applied. A high efficiency value corresponds to a network of sites that complement each other very well (have few points in common).

Flexibility: a good manager needs to have several alternative scenarios on hand when drawing up his network of reserves, should the "greenest" scenario not be acceptable from a political, social or economic point of view.

Irreplaceability: when the manager proposes several scenarios for the creation of reserves, certain sites are mentioned in almost all the proposals with a greater or lesser order of priority. The more a site is proposed, the greater its "irreplaceability"; in other words, it contributes more than the other sites to the aim of creating the reserves. It is therefore advisable to choose, from among the scenarios proposed, only those containing all the "irreplaceable" sites (or at least a good proportion of them).

On the other hand, there are two approaches that can be used to improve or extend a reserve network: the gap approach and the hotspot approach.

The gaps approach to selecting potential sites for conservation works by superimposing maps, each showing a very specific feature of the area under study. In this example, a map of plant communities, another of animal communities and a third showing existing reserves are superimposed. Superimposing the maps shows the 'gaps' in land protection according to the characteristics studied. This method is particularly effective for highlighting ecosystem protection needs. Such an approach places particular emphasis on the question of complementarity (Scott et al. 1991; Primack, 1993; cited by Lavoie, 1998).

The hotspots approach prioritises, when selecting sites for conservation, areas where biological diversity and endemism are very high and where protective action is urgent (imminent destruction). Using this method on a global scale, we can identify 12 hotspots in tropical forests (A) and 8 hotspots elsewhere in the world (B). This approach emphasises the issue of irreplaceability. On the other hand, it does not give enough weight, at least on a global scale, to the question of complementarity (Myers 1988, 1991; Primack, 1993; Schmidt 1996; quoted by Lavoie, 1998).

Reserve design

When a disturbance occurs in a balanced ecological system, it eventually returns to its original state. From a conservation point of view, this means that a natural unit can be maintained in a reserve as long as it is protected from human influence. In thc non-equilibrium model, the same disturbance can completely change the state of the ecological system in question, with no return to the original state. This model implies that a nature reserve does not function in isolation from its immediate environment and that it can be greatly and irreversibly influenced by the disturbances that affect it (Meffe and Carroll 1994; cited by Lavoie, 1998).

Reserve size

Large nature reserves contain larger populations of each species than smaller reserves. Only very large nature reserves can contain viable populations of large vertebrates.

In North America, Meffe and Carroll (1994; cited by Lavoie, 1998) have observed that small pares (1,000 km^2) are not large enough to support populations of large vertebrates, making extinction of the latter more likely. In fact, most of today's parks cannot support populations of large mammals on their own. It is estimated that only between 0 and 22% of the world's pares are capable of supporting large mammal carnivores (10-100 kg) over a period of more than 100 years, and that no park can support these carnivores over a period of more than 1,000 years.

Newmark (1996) observed Tanzania's national parks (ANP: Arusha; KNP: Kilimanjaro; LMNP: Lake Manyara; RNP: Ruaha; SNP/NCA: Serengeti and Ngorongoro; TNP: Tarangire) and the use of land adjacent to these parks. Six species of large mammals have disappeared from small parks (2,600 km2) (ANP, KNP, LMNP and TNP) since their creation between 35 and 83 years ago, whereas no species has disappeared from the very large Tanzanian parks (RNP and SNP/NCA: 13,000 to 23,000 km2) during their 50 years of existence.

Form of a reserve

The linear reserve does not really have a core zone, i.e. a zone located at a sufficient distance from the boundaries of the reserve so that it is not influenced by disturbing agents from outside. On the other hand, the circular reserve, which has the same surface area as the linear reserve, has a genuine core zone. Species typical of core areas can therefore survive in such a reserve, which is not the case in the linear reserve. It should be noted, however, that the linear reserve will be little affected by the loss of part of its territory, unlike the circular reserve, which will lose a substantial part of its core zone.

Area of land to be preserved in its natural state

The Brundtland Report was modest. It recommended that each nation should set aside 12% of its territory as protected areas (N.B. the DRC preserves barely 10% of its territory and the target is 15%). The 12% proposal has no scientific basis. There is therefore no reason to believe that by preserving 12% of the territory the conservation objectives will be achieved, when development is continuing intensively in the peripheral areas.

A limited number of studies have attempted to determine what space represents all the ecosystem types or species of a region in protected areas. In

1 Overall, Noss (1996; cited by Lavoie, 1998) estimates that between 25% and 75% (on average 50%) of the surface area of a region needs to be protected in order to achieve conservation objectives; here, protection >> does not mean non-development.

Two models of natural resource management

Two models emerged during the 60s and 70s: the exclusive model and the inclusive

model.
The exclusive model advocates state intervention in the management of natural resources. The proponents of the exclusive model, Hardin and Ophulus (1968; cited by Ofoumon, 1995), were opposed to joint management of natural resources. They argued that natural resources would deteriorate if individuals cooperated in their management. Individuals are reluctant to preserve natural resources; there must be a constraint for individuals to safeguard natural resources. It is the State that is capable of administering a sufficient dose of constraint. If ruin is to be banished, in a world prone to destruction, the population must be repressed by an external force that takes no account of their psychological state. To fulfil such a mission, we need an iron government, perhaps even a military government, which could do everything in its power to bring the ecological problems under control. The exclusive model based on protectionism has been applied in the USA and most of the national parks in sub-Saharan Africa have been created according to this model (Wells et al., 1992, Newmark 2000; Kramer et al., 1998).
Unfortunately, it is becoming increasingly clear that this model alone cannot guarantee the long-term conservation of biodiversity. This is particularly true in Africa, where most of the people living in enclave zones around protected areas are poor and exploit natural resources as a basis for survival (Haeckel, 1999).
The inclusive model advocates participation in natural resource management. Protected areas will prosper if and only if they are supported by the public, the private sector and government agencies (McNeely, 1995). Since the mid-1980s, the inclusive model has been applied in developing countries. Although its application dates back to the 1950s, the biosphere reserves of UNESCO's Man and the Biosphere (MAB) programme are the first experiences of systematic application of the inclusive model (Newmark, 2000; Kramer et al., 1998; Martino, 2001). Integrated conservation-development projects that combine the development of local populations with the conservation of natural resources are approaches that illustrate the application of the inclusive model.
Depending on the context, each of these two models is relevant to the conservation of natural resources. Generally speaking, the exclusive model is appropriate for preserving natural resources in the wild and in areas where armed poachers operate and where local populations are not involved in the destruction of natural resources. The inclusive model is appropriate in the case of inhabited protected areas and those whose protection affects local living conditions. Application of the inclusive model does not exclude police repression of poaching and any other form of fraud. The success of protected area management activities often depends on a combination of the two models (Garrity et al., 2000; Kasereka, 2003).

Integrating protected areas into regional planning

There is no point in endeavouring to conserve areas that are important for protecting biodiversity, which is itself essential to human survival, if at the same time we allow the destruction of natural resources outside these areas to continue. The amplifying feedback loop that is set in motion could eventually lead to the disappearance of these areas dedicated to nature conservation. It should also be noted that although protected areas are the main nature conservation strategy, they represent only 3.2% of the Earth's surface (Lavoie, 1998). There is no guarantee that these areas alone, even if kept intact, will suffice to safeguard biological diversity.

"A national park, a biosphere reserve or any protected area must be considered as a system whose environment is its territory. It is the riparian zone of the park that is the most sensitive; it is there that the populations that exert pressure on the protected area are to be found; it is this interface that is the scene of events (attacks on livestock by wild animals, devastation of fields, illegal occupation, poaching, etc.) that can degenerate into conflicts between the populations and the authorities responsible for the protected area. For these reasons, all protected areas must be considered in the context of land-use planning" (Maldague, 1997).

The planning of protected areas in their environment requires global and regional planning. In this context, they must be considered as a form of land use which is justified from the ecological, economic, psychosocial and cultural points of view. Land-use planning of this kind must lead, on the scale of a given area, to the rational use of resources, to the optimum organisation of human settlement in peripheral areas and to the creation of a quality environment which promotes the full development of individuals and the equilibrium of the community.

The model proceeds by drawing up development plans that cover the use of land, the requirements for developing the area's resources, the organisation of facilities and infrastructure, the nature and use of buildings, the services to be provided, etc. These plans will take concrete form in the form of development projects. These plans will take the form of development projects.

X. *Some protected area management activities*

There are several activities involved in managing a protected area. These include law enforcement, planning, tourism, research and monitoring, environmental education and rural development.

a. Applying the law

Protected areas are subject to human pressure. Poaching and habitat destruction are some of the most common problems, especially in developing countries. Law enforcement consists of recording and punishing offences. Surveillance patrols are organised by park rangers or forestry officers, who are usually armed. In the United States, the first managers of Yellowstone National Park were military personnel. In the Democratic Republic of Congo, the park rangers are paramilitary and carry rifles while carrying out their duties, while the conservators are judicial police officers.

b. Development

The activities involved in developing a protected area consist of zoning, delimitation, marking out tracks and various constructions. Zoning consists of defining and delimiting portions of space for different uses. Different categories of zones in a national park have been identified by Maldague (1990).

- Integral protection zone. There is no place for tourism, recreation or education in this zone. Even scientific research is subject to restrictions, such as access to the area and the handling and collection of biodiversity elements.
- Tourist use zone (recreational and educational). Tourists are allowed in this zone but their activity must be limited to observation. They are not allowed to pick anything or

collect any objects.

- Corridors. It may happen that tourist observation zones are separated by integral protection zones. In such cases it may be justified to provide a corridor linking different tourist use areas.
- Tourist use area (reception). This category covers visitor reception areas, entry and exit points, accommodation (hotels, guest houses, lodges, refuges, etc.), educational facilities (education centres, museums, etc.), etc.
- Zones reserved for research, surveillance and park management. This zone includes the infrastructure required for research and the smooth running of the park: weather station, airstrip, field laboratory, observation posts, patrol post, surveillance runways, communications and first aid post, garages, shops and warehouses.
- Buffer zone. An area on the periphery where land use is restricted in order to form a protective belt for the park while providing benefits to neighbouring rural communities (Martino, 2001).

c. Tourism

It is widely recognised that protected areas bring significant economic benefits to the country and to rural communities. In Kenya, a third of the US$ 125 million annual income from tourism comes from the seven most visited national parks. << tourism in a protected area consists of welcoming tourists, guiding them in their observations and offering them the necessary services, such as accommodation and catering.

d. Research and monitoring

The sustainable conservation of resources and the development of national parks require improved knowledge of the structure, revolution and functioning of ecosystems, the regulation of animal populations, the ecological and ethological relationships in these ecosystems, the vegetation, etc. Research is the key to accomplishing these complex tasks. In the Democratic Republic of Congo, research is the fundamental objective of the Institut Congolais pour la Conservation de la Nature (Kabemba, 1990).

Monitoring is the regular, standardised measurement and reporting of an activity or parameter over time (Hillman, 2001). Biomonitoring consists of monitoring the state of biocenoses. Law enforcement monitoring is a standardised method of tracking surveillance activities and threats to the protected area (e.g. poaching, occupation, habitat destruction) and the consolidation and use of this information by managers.

e. Environmental education

Environmental education consists in making the public and decision-makers aware of the interdependence between the environment and the economic development of a region, and in giving people the tools they need to deal with the environmental problems they face and to avoid creating new ones (ICCN, 2000). Using the discoveries of science and technology, education provides the knowledge, skills, attitudes and commitment needed to protect and improve the environment.

f. Rural development

Applying the inclusive model and integrating protected areas into land-use planning requires development activities to be carried out for the benefit of local communities. Protected areas that apply this approach pass on part of their tourism revenue to local communities, usually through structures set up by the beneficiaries. Conservation aid agencies in developing countries include support for local development among their priorities. In Kenya (Amboseli National Park), several members of local communities receive study grants for their children. In

In the Democratic Republic of Congo, some inhabitants of the hinterland of the Kahuzi-Biega National Park have benefited from the construction of schools and health centres, the improvement of roads and bridges, and support for agriculture, livestock farming and schooling for children (Kasereka, 2003).

Y. Nature conservation in the DRCongo

a. CONSERVATION POLICY

In terms of biodiversity, the Congo is the richest country in Africa and one of the richest in the world (STUART and ADAMS 1990; cited by ICCN, 2000). Its vegetation is dominated by lowland tropical forests and floodplain forests, which cover much of the country's central basin. Towards the north and south, this tropical forest changes to savannah and open forest. The east of the country is dominated by mountains that form the edge of the great African rift. Of its surface area of 2,345,409 km^2 , 45% is covered by closed forests. Of the 17 different Phytophtora in the Afrotropical biogeographical region, 7 are represented in Congo. This represents an exceptionally wide range of ecosystems and habitats: dense humid forests, wooded savannahs, grassy and shrubby savannahs, Afro-montane forests, mangroves, papyrus, peat bogs, marshes and flooded savannahs. 52% of the national territory is covered by forest, i.e. 47% of the surface area of the African tropical forest. In terms of species diversity, the DRC ranks first among African countries for several taxonomic groups: 409 species of mammals, 1,086 species of birds. 1,069 species of fish, 152 species of snake, 20 species of cameleon, 15 species of tortoise and freshwater turtle, 105 species of aquatic mollusc, 167 species of aquatic chironomid, more than 1,000 species of butterfly and more than 10,000 species of angiosperm, 3,000 of which are endemic.

The importance of this wealth was recognised by both the colonial authorities and the government of the Democratic Republic of Congo. Protected areas are the main strategy for conserving biodiversity in the DRC. For the purposes of conserving fauna, flora, soil, water and the natural environment in general, any part of the Republic may be set aside as an integral reserve (Ordonnance-loi no. 69-041). More than 10% of the territory is set aside as national parks, hunting grounds, wildlife reserves and various classified areas (World Bank, 1993; ICCN, 2003).

The strategy based on protected areas remains the best option for the future conservation of floral and faunal biodiversity in the DRC, because of the many economic and management problems facing the country (Hart and Hall, 1996).

LEGAL FRAMEWORK

Prior to independence, protected areas were established by Royal Decree, covered by the General Decree on Conservation of 1937. On 22 August 1969, Ordinance-Law No. 69-041 on nature conservation was adopted, confirming the existing protected areas and creating the Institut Congolais pour la Conservation de la Nature as the management body. This legislation was adapted by Decree No. 244 of 21 February 1972, Decree No. 023 of

22 July 1975 and Ordinance No. 73190 of 15 May 1978. The restrictions laid down for national parks generally meet the IUCN criteria for category II (national parks) in the classification of protected areas. Thus, all acts likely to cause direct damage to the components of the ecosystem are prohibited (art.5). However, the organisation of tourism and travel essential to the economic development of the population are permitted. The legislation does not provide for the creation of buffer zones as such, but gives the park authorities the power to record offences up to 50 km beyond the boundaries. The Ordinance also stipulates that national parks are created by presidential decree on a proposal from the government after a vote in parliament and without any prior investigation procedure.

Three other laws are of particular importance for nature conservation Land legislation, Law No. 80.008 of 18 July 1980 adapting the Bakajika Law of 20 July 1973, determining that the soil and subsoil belong to the State. The Hunting Law, Law No. 82.002 of 28 May 1982, determining which animals are totally protected as well as the different hunting permits, hunting seasons and prohibited hunting techniques. A decision of the Executive Council in 1984 suspended hunting throughout the country.

The 1986 "Guide de l'exploitant forestier", based on the Forestry Law of 11 April 1949 and land law, obliges the beneficiary of a forestry concession (large or small) to develop the land.

A new law promoting participatory management of protected areas is currently being drafted.

INSTITUTIONAL FRAMEWORK

Following the Conference on the Protection of Fauna and Flora in Africa, held in London in 1933, the colonial authorities established the Institute of National Parks as an autonomous body within the Department of Agriculture in 1934. After independence, the national parks were administered directly by the Department of Agriculture. Since 1969, the ICCN has been responsible for the management of national parks and hunting areas, including the protection of flora and fauna, scientific research and tourism in accordance with the fundamental principles of nature conservation. Since 1969, ICCN has been a public enterprise under the Ministry of the Environment, Nature Conservation and Tourism. It is headed by a President Delegate General appointed by the President of the Republic and has its headquarters in Kinshasa with 3 different departments: the administrative and financial department, the technical department and the research department.

SOME OF THE ICCN'S PARTNER ORGANISATIONS

- AWF, African Wildlife Foundation
- CI, Conservation International
- DFGF/E, Dian Fossey Gorilla Fund Europe
- DFGF/I, Dian Fossey Gorilla Fund International
- FFI, Fauna and Flora International
- FZS, Frankfurt Zoological Society
- GIC, Gilman International Conservation
- GIZ, German Cooperation for International Development
- JGI, Jane Goodall Institute

- IGCP, International Gorilla Conservation Programme
- IUCN, International Union for Conservation of Nature
- UNESCO, United Nations Educational, Scientific and Cultural Organization
- USAID, United States Agency for International Development
- WCS, Wildlife Conservation Society
- WWF, World Wide Fund for Nature
- ZSL, Zoological Society of London

PROTECTED AREAS IN THE DEMOCRATIC REPUBLIC OF CONGO

There are 6 main types of protected area in the Democratic Republic of Congo: national parks, hunting areas, biosphere reserves, forest reserves, zoological and botanical gardens and protected areas.

Their distribution in terms of surface area, and for those for which data are available, is as follows:

Table 1: Importance of protected areas in the Democratic Republic of Congo

TYPE	NUMBER	SUPERF. APPROXIMATIVE	% OF COUNTRY
National parks - Operational - In project	 8 4	 8.491.000 2.244.625	 3,6 0,9
Hunting grounds	57	10.954.266	4,6
Biosphere Reserves	3	267.414	0,1
Forest reserves	117	517.169	0,2
Zoological and Botanical Gardens	3 3	3.000	0,0+
Backup sectors Reforestation sites Nature reserves	 1	 112.000 36.000	 0,0+ 0,0+
TOTAL		**22.655.474**	**9,6%**

Source: ICCN and DGNR compilation (1994)

N.B. Ramsar sites

These are wetlands of international importance. The Convention on Wetlands, signed in Ramsar, Iran, in 1971, is an intergovernmental treaty that provides a framework for national action and international cooperation for the conservation and wise use of wetlands and their resources. In the DRC, three sites are on the Ramsar list. These are Virunga National Park (1996), Mangroves Park (1996) and Ngiri-Tumba-Mai Ndombe (2008).

The aim of the Democratic Republic of Congo is to increase this area to 12-15% of the national surface area, taking into account the different natural ecosystems that reflect the biological diversity specific to the Congo. Thirteen sites of national and international interest and 4 sites of local interest have been identified by the IUCN to form part of the network of protected areas.

National Parks

In colonial times, each protected area was established by Royal Decree on the basis of a general decree signed in 1937. The national parks were all established by specific

ordinance-laws confirming their status as "Integrated Nature Reserves", although certain scientific and tourist activities are now authorised.
Management of the parks is the responsibility of the Institut Congolais pour la Conservation de la Nature (ICCN) in accordance with Ordinance-Law no. 75-023 of 22 July 1975.
There are currently 7 national parks:
Virunga Park, Garamba Park, Kahuzi-Biega Park, Salonga Park, Upemba Park, Kundelungu Park and Mai'ko Park. The first four are on UNESCO's list of World Heritage sites. Salonga and Virunga are home to 2 Ramsar sites: Ngiri-Ntumba- Mai Ndombe and Lac Edouard. In addition to these 7 parks, there is the Parc de la N'sele (3,000 t), made up of a part where the animals live in freedom and another where they are kept in captivity (Zoo of 17 Ha).
A project to create three new national parks has been envisaged and included:

- Okapi National Park (1,372,625 Ha): designed to protect the Okapi, a rare species endemic to the Congo;
- The Mondjo National Park (812,000 hectares) will protect the degraded dense dry forest and the herds of elephants it shelters;
- the Moanda Marine Park (60,000 Ha), to protect the special biotope of the Atlantic coast (Mangroves with Mangrove trees), (Ramsar site).

In addition, the ICCN is considering proposing that the Swa-Kibula Reserve be declared a park, in order to protect the characteristic ecosystem of the country's southern savannah strip, with its typical forest galleries.

Hunting grounds

Their management used to be the responsibility of the Ministry's Direction de Gestion de Ressources Naturelles Renouvelables. It was transferred to ICCN by ministerial decree no. 36/DECNT/BCE/78 of 13 July 1978.
Of the 57 areas identified, 19 are actually operational or functional. The others have been virtually abandoned for material and financial reasons.

Biosphere Reserves

From a legal point of view, these areas are in principle governed by FUNESCO concepts. In addition, the latter issues permits or certificates on an ad hoc basis. There is no legal act giving concrete form to UNESCO's action. However, management and supervision are carried out by the National Secretariat of the MAB Programme in Congo, which is attached to the Ministry of the Environment. Of the three Reserves that existed at the outset (Yangambi, Luki and Lufira), only the Luki Reserve (33,000 ha) is still being monitored.

Forest Reserves

The management of these Reserves is the responsibility of the Directorate for the Management of Natural and Renewable Resources (DGRNR) attached to the Ministry of the Environment, Nature Conservation and Tourism. In principle, there are no legal texts governing their management. Their objectives are to protect the natural forest, promote silvicultural research and enrich the natural forest with a view to promoting the production of timber or firewood.

The Zoological and Botanical Gardens

This is a type of ex-situ conservation of flora and fauna. These gardens are under the responsibility of the Institut des Jardins Zoologiques et Botaniques du Congo (UZBC), established by order no. 78-215 of 5 May 1978.

There are currently 4 zoological gardens in Kinshasa (2), Kisangani (1) and Lubumbashi (1) and 3 botanical gardens in Kisantu (Bas-Congo), Eala (Equateur) and Kinshasa (Parc de la Revolution).

Safeguard sectors

Act no. 75-024 of 22 July 1975 defines the principle of their creation and sets out their legal and organisational framework.

Under the terms of this law, any area not subject to legally recognised human easements and of particular interest from the point of view of conservation, restoration or development must be safeguarded.

Although article 3 of the aforementioned law stipulates that the body responsible for managing these Sectors must be designated by the order establishing it, nothing has yet been done in this regard.

Annex: List of international agreements with an impact on biodiversity and of which the DRC is a member

NAME OF AGREEMENT		COUNTRY OR TOWN OF ADOPTION	DATE OF SIGNATURE
1	Phytosanitary Convention for Africa south of the Sahara	London (England) 29 July 1954	21 September 1962
2	Cooperation agreement on quarantine and plant protection against pests and diseases	Sofia (Bulgaria) 14 Dec. 1959	
3	African Convention on the Conservation of Nature and Natural Resources	Algiers (Algeria) 15 September 1968	13 November 1976
4	Convention on Wetlands of International Importance particularly as waterfowl or Ramsar habitats	Ramsar (Iran) 2 february 1971	15 September 1994
5	Convention concerning the Protection of the World Cultural and Natural Heritage	Paris (France) 23 November 1972.	17 december 1975
6	Convention on the Conservation of Migratory Species of Wild Animals wildlife (Bonn)	Bonn (Germany) 23 June 1979	5 September 1994
7	Vienna Convention on the Protection of the Ozone Layer; London and Montreal Protocols	Montreal (Canada) 22 March 1985	15 September 1994

8	framework convention on climate change	Rio de Janeiro (Brazil) 4 June 1992	8 december 1994
9	Convention on Diversity Organic	Rio de Janeiro (Brazil) 4 June 1992	15 September 1994
10	Convention on the Transboundary Shipment of Hazardous Wastes and their Management (Bamako Convention)	Bamako (Mali)	15 September 1994
11	International Tropical Timber Agreement	Geneva (Switzerland)	20 November 1990
12	United Nations Convention on the Law of the Sea	MontegoBay (Jamaica) 10 december 1982	17 february 1989
13	Convention on the Transboundary Shipment of Hazardous Wastes and their Treatment (Basel Convention)	Basel (Switzerland) 22 March 1989	15 September 1994
14	Convention on the Conservation of European Wildlife and Natural Habitats	London (England) 14 January	
15	Phytosanitary Convention for Africa	Kinshasa (Congo) 13 September 1967	
16	International Plant Protection Convention	Rome (Italy) 6 december 1951	16 September 1975
17	Convention on the Prohibition of the Development, Production and Stockpiling of Bacteriological (Biological) and Toxin Weapons and on their Destruction	Washington (USA) 10 April 1972	10 september 1975
18	Convention Concerning the Protection of the World Cultural and Natural Heritage	Paris (France) 23 November 1972	17 december 1972
19	Convention on the Prevention of Marine Pollution by Dumping of Wastes and Other Matter	London 29 december 1972	17 October 1975
20	Convention on International Trade in Endangered Species of Wild Fauna and Flora (CITES)	Washington (USA) 3 March 1973	18 October 1976
21	Convention on the Conservation of Migratory Species of Wild Animals belonging to wild fauna	Bonn (Germany) 23 June 1973	15 September 1994
22	Convention to combat desertification and drought	Paris (France) 17 October 1995	11 September 1997

CHAPTER II

STUDY ENVIRONMENT AND METHODOLOGY

II.1 STUDY ENVIRONMENT

a) Geographical location

It is a village located in the KAGABI group, KABARE Territory, South Kivu Province in the Democratic Republic of Congo. Administratively, it is run by the village chief, Mr LUSARI. The village MBIZA is subdivided into three sub-villages:

S MBIZA PREMIER: Includes the BUHIRWA and KAJEJE sites.

S MBIZA SECOND: Includes the KABUGE and IKUMBANI sites.

S MBIZA CENTRE: Includes the sites of KIYABAGALA , KABOMBO , NYAMUKUMBWA and MURUNDU.

The climate is temperate and high altitude, with average temperatures varying between 15 and 25°. The region has two seasons: the rainy season, which lasts 8 to 9 months, the long rainy season begins in August and ends in December, and the short rainy season begins in February and ends in May. The long dry season, which lasts 3 to 7 months, starts in June and lasts until December, while the short dry season starts at the beginning of January and lasts until the end of the same month.

b) Soil and vegetation

This aspect is of greater concern in this research. The MBIZA village has clay-limestone soils that are favourable for farming. In addition, there is a lack of agricultural support in the form of extension services, seed certification and a shortage of chemical and organic fertilisers. This has led to progressive soil degradation, which in turn has caused infertility. As far as vegetation is concerned, the village of MBIZA has herbaceous and woody plants on both sides that characterise its landscape. It should be noted here that the relief is generally mountainous. The main crops grown in MBIZA are maize, beans and manioc. These crops, which used to give considerable yields in this area and contributed to food security, have seen their yields fall in recent years as a result of the problem of infertile soils. At present, the agricultural soils in this area are no longer suitable for these different crops as they were in the past, having lost their quality.

Soil infertility is currently a major problem in the DRC in general and in MBIZA in particular, and in the KAGABI group it is due to the following factors:

- The lack of improved seeds;
- Non-rotation of crops ;
- The use of traditional methods of soil fertilisation ;
- Farmers' failure to comply with GIFS :
- Growing on slopes.

These different causes of soil infertility in the KAMURONZA groupement are explained in the pages below as follows:

4- The lack of improved seeds

As we can see here, improved seeds are of vital importance to the stability of agricultural production, given their resistance to disease and pests. However, as can be seen in the province of South Kivu, there is a virtual shortage of improved seeds that can rejuvenate agricultural soils and contribute to their fertility. The non-availability of improved seeds can be explained by the fact that the agricultural sector receives a very small percentage of the national budget. But poverty also prevents farmers from being able to obtain them in areas where they are available. This is why, in our field of study, we continue to use local varieties that are not resistant to diseases and pests. In the long term, this leads to infertility on the land.

5- Non-rotation of crops :

In view of the fact that soil impoverishment and an increase in diseases and pests were observed when farmers grew the same crops in the same field for several years in succession, crop rotation was recommended.

6- The use of traditional methods of soil fertilisation :

Continual soil deterioration is a major obstacle to increasing yields. When farmers cultivate the same fields, season after season, and cannot afford to replace the soil nutrients absorbed by their crops, the soil is in fact depleted of its substance. Precarious cultivation techniques and practices such as bush clearing, no rotation and slope cultivation contribute to the loss of soil fertility.

4- Farmers' failure to comply with GIFS :

ISFM is a prerequisite proposed by agronomy specialists to help ensure the fertility of agricultural soils.

4- Growing on slopes

Some slopes are steeper than others, and the steeper the slope, the harder it is to work. When cultivating on a slope, when it rains heavily, the water can wash the good soil straight down the slope to the bottom of the hill. This washes away the topsoil, which contains most of the nutrients and organic matter in the soil. When farmers cultivate across the slope, the damage is significant. As long as there is a furrow, it will form a small gully to retain the water and prevent it from running down the hill and eroding the soil. For farmers cultivating on the higher slopes, it is difficult for their land to be drained, which has an impact on the infertility of the soil. This is the remarkable situation in the village of MBIZA, where some farmers are suffering the consequences of growing crops on slopes.

4- Failure to respect the agricultural calendar :

This is not the case in MBIZA Village, where farmers are overexploiting the soil month after month due to famine, to the point of depleting the soil to a major degree. What's more, by not leaving the land fallow, the soil loses its buffering capacity and becomes infertile.

II .2 .WORK METHODOLOGY

1.3.1. Data collection procedure

To gather the information needed for this study, we went through several stages, including pre-survey, documentation, drafting of the survey questionnaire and field surveys.

In order to find out the actual situation regarding the problem addressed in this research, that of soil fertility management in the MBIZA village, we felt it was important to carry out a preliminary survey. This being the case, we arrived at different sites in the MBIZA village where this problem is at its worst and observed the situation in the field before going on to prepare the work. After the preliminary survey, we moved on to the documentation stage, to enable us to gain access to the necessary data contained in various scientific documents, namely books, reports, articles, dissertations and theses dealing with our research topic, in order to enrich it. To find out about these documents, we consulted websites and also visited a number of local libraries. Once our research was complete, we set about drafting the introductory section of the study and the survey questionnaire. The questionnaire was drawn up on the basis of the hypotheses and objectives of the work. The questionnaire was addressed to the peasant farmers of MBIZA Village. Once we had completed the questionnaire, we multiplied it according to the size of the sample and went out into the field to carry out the surveys. For the field surveys, we travelled through almost the entire village of MBIZA, carrying our survey questionnaires with us and administering them to the peasant farmers. Once we had completed our surveys, we interviewed them about the questions on the questionnaires. The farmers gave us their views and opinions on the problems of soil fertility management in their area. We transcribed the responses from the surveys onto the questionnaire forms, as the vast majority of respondents were illiterate. To save time during the surveys, we also recorded the opinions of the respondents in terms of their answers to the questions asked.

Once we had finished interviewing the respondents, we visited the various arable fields to find out what was really happening on the ground.

1.3.2. Sampling

To determine the sample size, we used the COCHRANG formula, quoted by Professor Alain BOUCHARD. This formula states that when the size of the universe is infinite, i.e. greater than 1000000 inhabitants, we take a sample of 96 people. When the size of the universe is finite, we apply the formula for corrected size, assuming a margin of error of 10%. This formula is as follows:

$NC = N \times \frac{n}{N+n}$ or: NC= Sample size

N=Size of population

n=Size of the universe

Given that the population of the Kagabi district is estimated at 43616, we apply the COCHRANG formula as follows:

$$NC = 43616 \times \frac{96}{43616+96} = \frac{4187136}{43712} = 95,78 \approx 96$$

Margin of error 10%: $96 \times \frac{10}{100} = 9,6 \approx 10$

96 individuals + 10 individuals = 106 people.

Hence our sample size of 106 individuals.

1.3.3. Statistical analysis

For statistical analysis and data processing, we used SPSS 21 software. Using this software, we encoded the data in Excel to produce high-quality tables illustrating the study.

CHAPTER III

PRESENTATION, ANALYSIS AND INTERPRETATION OF RESULTS

Table 1 below shows that the majority of people surveyed were aged between 15 and 25, 51 out of 106 (48.1%), while the minority were aged 60 and over, 3 out of 106 (2.8%). This is justified by the fact that it was the young farmers who were permanently available during our surveys and above all that it was they who were active in the fields and it was the old farmers who gave us information about the history of agriculture in the area.

For the marital status variable, the table shows that 76 of the 106 people interviewed were married, i.e. 71.7%, while 7 of the 106 interviewed were divorced, i.e. 6.6%. This is justified by the fact that, in rural areas, marriage is contracted too early and, above all, that villagers fear divorce because of the costly demands it entails.

With regard to the variable level of education, the results of this table show that 41 people out of 106, i.e. 38.7%, have secondary education, 36 people out of 106, i.e. 34.0%, have primary education, 18 people out of 106, i.e. 17.0%, have no level of education, i.e. are illiterate, and 11 people out of 106, i.e. 10.4%, have university education. With regard to these levels of education, it can be seen that most of the farmers in MBIZA Village do not have a high level of education because many left school early as a result of one circumstance or another. With regard to the size of the households surveyed, we found that the vast majority of respondents had households of between 6 and 10 people, with 40 people out of the 106 surveyed, or 37.7%. For others, their household size is between 1 and 5 people, 38 people out of 106, i.e. 35.9%, and finally we have 28 people out of 106, i.e. 26.4%, whose household size is 11 people or more. We can see that household size is high in MBIZA village, a situation that can be explained by the high birth rate in rural areas.

Table 1. Socio-demographic characteristics

№	Variables	Procedure	Workforce	Percentages (%)
1	GENRE	- WOMEN - MALE	- 47 - 59	- 44.3 - 55.7
	AGE	- 15-30 - 31-45 - 46-60 - 60 and over	- 51 - 27 - 25 - 3	- 48.1 - 25.5 - 23.6 - 2.8
2	CIVIL STATUS	- CELIBATORY - DIVORCE - MARIE (E)	- 23 - 7 - 76	- 21.7 - 6.6 - 71.7
3	LEVEL STUDY	- PRIMARY - WITHOUT - SECONDARY - UNIVERSITY	- 36 - 18 - 41 - 11	- 34.0 - 17.0 - 38.7 - 10.4
4	SIZE MENAGE	- 1 a 5 - 6 a io	- 38 - 40	- 35.9 - 37.7

		- 11 and over	- 28	- 26.4

Table 2 below shows that the MBIZA Village surveys confirm the possession of land for cultivation.

From this table, we can see that the vast majority of people surveyed have fields for food crops, 94 out of 106, or 88.8%. The minority did not own any arable land in our study area. They represent 11.3% of our sample. For those who do, the reason is that they either inherited the fields from their fathers, bought them or rented the land to farm. For those who do not, they lack the means. In the light of this table, it is clear that the size of the fields to be farmed differs from one farmer to another. As a result, as the results of this table show, the majority of people interviewed have fields of less than one hectare, 47 people out of the 106 surveyed, or 44.3%. It is only farmers with the means to do so who are able to acquire large tracts of land.

Table 2. Ownership of arable land

Variables	Procedure	Workforce	Percentages (%)
Land ownership arable land	YES	94	88,7
	NO	12	11,3
Extent of land cultivable	1ha	28	26.4
	2ha	12	11.3
	3ha	4	3.8
	4ha	3	2.8
	Less than 1 ha	47	44.3

Table 3 below shows that the arable land of farmers in MBIZA Village is not fertile. This was stated by 75 people out of the 106 surveyed, i.e. 46.2%. This is justified by the many problems linked to the fertility of the agricultural soils and their management, which we have elucidated in the preceding pages. This infertility has a significant impact on yields. In the light of the results shown in this table, it can be seen that with regard to the situation of the arable land in relation to the slope, for many farmers in the village of MBIZA their land is located on the average slope, as stated in 41 out of 106 surveys, i.e. 38.7%. The farmers explained that the higher slopes are less fertile because when the soil is washed away, all the plant nutrients tend to sink to the bottom, so they are content to cultivate on the lower slopes where the land is easily irrigated. These surveys represent 23 people out of 106, i.e. 21.7%. Then we have the respondents whose land is located in the lowlands, which are favourable for farming, 22 people out of 106, i.e. 20.8%. For the remaining respondents, their land is located in the savannahs (11 out of 106 respondents, i.e. 10.4%) and on the high slopes (9 out of 106 respondents, i.e. 8.4%).

Table 3. Confirmation of the fertility of arable land

Variables	Procedure	Workforce	Percentages (%)
Confirmation on the soil fertility arable land	No answer	12	11.3
	GUI	75	46.2

	NO	19	42.5
Situation on the ground cultivable in relation to a the slope	The shallows	22	20.8
	The savannahs	11	10.4
	Gentle slope	23	21.7
	High slope	9	8.5
	Average slope	41	38.7

Table 4 below shows that land infertility in the KAGABI group has many causes. As the results in this table show, the main causes of this problem in this area are respectively the lack of improved seeds identified by 28 people out of 106, i.e. 26.4%, the non-rotation of crops identified by 18 people out of 106 interviewed, i.e. 17.0%, non-compliance with the ISFM identified by 16 people out of 106, i.e. 15.1%, and the use of traditional methods such as monoculture, unimproved soil fertilisation inputs, etc. identified by 16 people out of 106, i.e. 15.1%. raised by 16 out of 106 people (15.1%).

Apart from these main causes, there are also secondary causes of soil infertility in MBIZA Village. These include soil leaching, reported by 1.9% of respondents, monoculture, reported by the same proportion of respondents, and many other causes, such as soil not being left to fallow and field ageing.

Table 4. Causes of soil infertility in MBIZA

Causes	**Workforce**	**Percentage**
Other	2	1.9
Soil leaching	2	1.9
the use of methods traditional soil fertilisation	16	15.1
Growing on slopes	12	11.3
Monoculture	2	1.9
Non-rotation of crops	18	17.0
The lack of improved seeds	28	26.4
non-compliance with ISFM (Integrated Soil Fertility Management)	16	15.1
Failure to respect the agricultural calendar	9	8.5
Over-exploitation of the field	1	0.9
Total	106	100.0

Table 5 below shows that traditional methods of fertilising the land are still used by farmers in MBIZA Village. As a result, the methods most commonly used by these farmers are slope cultivation (35.8% of respondents), monoculture (27.4% of respondents), the use of unimproved inputs and slash-and-burn agriculture (17.9% of respondents). The farmers' reasons for using these techniques are maintenance work and enriching the soil with nutrients, but in reality these techniques damage the soil and do not guarantee the expected fertility. Farmers in MBIZA use a number of modern techniques to fertilise their land. The vast majority of these farmers (42 out of 106 surveyed, i.e. 39.6%) state that they use a combination of crops, while a minority (17 out of 106, i.e. 16.0%) use chemical fertilisers. It should be noted that, as we have seen at

In MBIZA, so-called modern techniques are practised to a small extent by peasant farmers, either because of a lack of support, a lack of information or a lack of mastery of one of the techniques applied. The soil is the site where agriculture is practised, and this has negative effects on the socio-economic conditions of the farmers.

Table 5. Traditional methods used to fertilise land

Variables	Procedure	Workforce	Percentages (%)
Methods traditional for the fertilisation of land	- Bush farming	19	17,9
	- Network planting	2	1,9
	- Use of non improvements	1 Q	17,0
		to	27,4
	- Monoculture	29	35,8
Methods modern for the soil fertility	Agroforestry	17	16.0
	Association of crops		QO A
		42	39.6
	Fallow land	30	28.3
	Use of fertilisers chemical	17	16.0

Table 6 below shows that the soil in the village of MBIZA is currently infertile, due to the serious problems that cause infertility. 86 of the 106 people surveyed, or 81.1%, said so. These results show that the three main crops in the village of MBIZA are sweet potatoes, cassava and beans. This is evidenced by the large number of surveys, 44 out of 106, or 41.5%. We can therefore say that the agricultural soils in this area are well adapted to these crops compared with others.

Soil infertility has a negative impact on agricultural yields in MBIZA Village, because crops cannot grow and adapt better on infertile soil. 80 people out of 106 surveyed, or 75.5%. Other respondents said that infertile soils had a very negative impact on crop yields, 17 out of 106 respondents, or 16.0%. We can therefore report that agricultural yields are currently low in MBIZA village as a result of soil infertility.

Table 6: Confirmation of soil infertility, details of main crops in MBIZA and influence of soil infertility on crop yields

Variables	Procedure	Workforce	Percentages (%)
Confinnationsur the infertility status of soils	No answer	3	2.8
	NO	17	16.0
	YES	86	81.1
The three cultures of the environment	No answers		
	Sugar cane, beans and sweet	3	2.8

	potatoes		
	Colocases, potatoes	2	1.9
	sweet and bean	2	1.9
	Colocases, potatoes	3	2.8
	soft and but	4	3.8
	Beans and manioc	5	4.7
	Beans, maize and manioc	1	0.9
	beans, manioc and amaranth	19	17.9
	Maize, beans and cassava	8	7.5
	But, cassava and potato	6	5.7
	sweet	4	3.8
	Cassava, beans and	2	1.9
	banana	44	41.5
	Cassava, beans, sugar cane	3	2.8
	sugar		
	Cassava, maize, groundnuts		
	sweet potatoes, manioc and beans Rice, manioc, maize		
Influence of soil infertility on agricultural yield in KAMURONZA	No Negatively Positive Very negative	3 80 6 17	2.8 75.5 5.7 16.0

Table 7 above shows that the current state of agricultural production in the village is low. This is justified by the fact that infertile soils are within the reach of farmers and have a negative impact on yields. 75 people out of the 106 surveyed (70.8%) said this. This is why it is important that mechanisms to restore soil fertility are put in place to overcome the situation in this area.

Table 7. Current state of agricultural production in MBIZA Village

Current state of production	**Workforce**	**Percentage**
No answer	4	3.8
Low	75	70,8
Average	21	19.8
very low	6	5.7
Total	106	100.0

Table 8 below shows that the average sweet potato yield varies from one farmer to another and according to the condition of the soil. The average yield for most farmers was 25 to 30 bags. However, this yield is too low to guarantee food security in this area. The average yield for these crops is between 25 and 30 bags, which, as our surveys have shown, is insufficient for all farmers. The average yield of maize is mediocre, as in the case of sweet potatoes in MBIZA village. It is slightly higher than that for sweet potatoes, reaching more than 25 bags per hectare. Despite this, the yield is low in relation to the number of inhabitants and the food needs to be covered.

As with the other two crops, bean production is low, varying between 1 and 5 bags, or 83% of the surveys.

Table 8. Average yields for sweet potatoes, maize and beans for farmers in MBIZA

Variables	**Procedure**	**Workforce**	**Percentages (%)**
Performance medium for sweet potatoes	No answer	22	20.8
	1 to 5 bags	19	17.9
	6 to 10 bags	35	33
	11 to 15 bags	18	17
	16 to 20 bags	3	2,8
	21 to 25 bags	4	3.7
	26 to 30 bags	5	4,7
Average yield for but	We don't know not	23	21.6
	1 to 5 bags	30	28.3
		30	28,3
	6 to 10 bags	17	16
	11 to 15 bags	3	2,8
	16 to 20 bags	2	1.9
	21 to 25 bags	1	0.9
	More than 25 bags		
Average yield for beans	We don't know	13	12,2
	1 to 5 bags	88	83
	6 to 10 bags	3	2,8
	11 to 15 bags	2	1,9

Table 9 below shows that the majority of respondents, 63 out of 106 (59.4%), confirm that agricultural production has a negative impact on the socio-economic and environmental conditions of farmers in the MBIZA group, while 43 out of 106 (40.6%) deny this. In reality, it is certain that agricultural production has a negative impact on the socio-economic conditions of farmers in this area, as production is low and cannot cover the social and economic needs of farmers.

Table 9: Impact of agricultural production on the socio-economic conditions of farmers in the village of MBIZA.

Impact	**Workforce**	**Percentage**
Negatively	43	40.6
Positive	63	59.4
Total	**106**	**100.0**

Table 10 below shows that 85 people106 (80.2%) confirm that NGOs working in the agricultural sector have already put in place strategies to promote good soil fertility management and increased agricultural production in MBIZA. The NGOs working in the agricultural sector are endeavouring to support farmers in the province of South Kivu in order to improve the agricultural activity on which the country depends.

Table 10. Confirmation of the strategies already put in place by agricultural NGOs for good soil fertility management and increased agricultural production in MBIZA.

Variable	**Workforce**	**Percentage**
NO	21	19.8
YES	85	80.2
Total	**106**	**100.0**

Table 11 below shows that NGOs use different strategies to contribute to good soil fertility management and increased agricultural production in MBIZA. The best strategies revealed by the surveys for this purpose are raising awareness of ISFM, fallowing and integrating improved seeds, as identified by 5.7% and 4.7% respectively.

Table ll. Strategies already put in place by NGOs

Strategies	**Workforce**	**Percentage**
No strategic	21	19.8
Application of GIFS	9	8.5
crop association, improved seed	4	3.8
improved seed distribution	5	4.7
Technical management	6	5.7
FAO	11	10.4
FAO, FEPACO	3	2.8
FAO, WHO	3	2.8
FAO, RESCU	2	1.9
FAO, RESCU, OFEPACO	2	1.9
FAO, VIFED	2	1.9
Seed integration, set-aside	3	2.8
Fallow land, Integration of improved seeds (FAO, SIDIP)	5	4.7
Improved seeds, fertilizer trees (FAO, RESCU)	2	1.9
MSF, WHO, MCP, ACF	4	3.8
WFP, FAO	4	3.8
PAM, RED CROSS	3	2.8
Planting shrubs	2	1.9
raising awareness of soil conservation techniques	3	2.8
GIFS awareness	6	5.7
SIDIPE	4	3.8
VIFEDE, FAO	2	1.9

Total	106	100.0

Table 12 below shows that the strategies implemented by agricultural NGOs have not been effective in promoting good soil fertility management. This was proven by 60 out of 106 respondents, i.e. 56.6%. The ineffectiveness of NGO strategies is justified by the fact that these NGOs operate irregularly and do not have the support of the Congolese government for this purpose. As for the wish to increase production and improve soil fertility in MBIZA, the vast majority of our respondents expressed the wish to increase agricultural production and improve soil fertility management in their area. 72 people out of the 106 surveyed, i.e. 69.9%, condemned this. This is justified by the fact that it should contribute to food security in the area and increase farmers' incomes.

Table 12. Confirmation of the effectiveness of NGO strategies and the desire of surveys to increase agricultural production and improve soil management in MBIZA.

Variables	**Procedure**	**Workforce**	**Percentages (%)**
Confining	No answer	21	19.8
I' efficacitedes	NO	60	56.6
NGO strategies	YES	25	23.6
Wishes for	No answer		19,8
the increase in	NO	21	12.3
agricultural production			
and	YES	13	67.9
Improving the			
soil fertility management		72	

Based on the results of Table 13, we note that 34 people out of 106, i.e. 32.2%, make no suggestions for good soil fertility management, 27 people out of 106, i.e. 25.5%, suggest raising awareness of new cultivation techniques, 20 people out of 106, i.e. 18.8%, suggest adopting new crop varieties, and 16 surveys out of 106, i.e. 7.5% per response, suggest applying ISFM for good soil fertility management.

Table 13: Suggestions for good soil fertility management

Proposals	**Workforce**	**Percentage**
No	34	32,1
Adoption of new crop varieties	20	18,8
Application of GIFS	8	7.5
Stop growing crops on slopes	1	0.9
Anti-erosive crops	2	1.9
Technical and financial support for farmers	8	7,5
Respecting the agricultural calendar	6	5.7
raising awareness of new cultivation techniques	27	25.5
Total	**106**	**100.0**

CORRELATION MATRIX

	Is the soil in your area in a good condition? infertility.	How does soil infertility affect crop yields in your area?	What is the current state of agricultural production in the infertile state of the soil for these three crops, in contrast to before?	What is the average yield for the th ree Main crops grown over the last 5 years in your area (Sweet potato)	What are the three main cultures in your environment?	What is the average yield for the three main crops grown over the last 5 years in your area (maize and manioc)?	What is the average yield for the three main crops over the last 5 years in your area (Beans)?	How does this agricultural production for crops affect your socio-economic life?
Coeff de corr floors your environment	1.000	.008	.296"	.350"	.216*	.176	.115	-.088
do they have Sig. (bilateral) infertility?		.938	.002	.000	.027	.071	.239	.370
N	106	106	106	106	106	106	106	106
How does this Coeff de correl infertilite des	.008	1.000	.166	.106	.191*	-.244*	-.038	.015
soil influence Sig. (bilaterally) on the yield of N	.938		.088	.282	.049	.012	.700	.876
thesecultures inyour environment	106	106	106	106	106	106	106	106
Correlation coefficient	.296"	.166	1.000	.348"	.115	.167	.153	-.049
What is Γ etat Sig. (bilaterale) dela agricultural N	.002	.088		.000	.242	.088	.118	.619
production currently at Γ etat d'infertilité des sols pour ces trois cultures contrasts à celle d'avant	106	106	106	106	106	106	106	106
What is the yield correlation coefficient?	.350"	.106	.348**	1.000	.117	.147	.029	.093
(bilaterale) three cultures	.000	.282	.000		.232	.132	.764	.343
main N corns over the last 5 years in your area (Sweet potato)	106	106	106	106	106	106	106	106
What are the correlation coefficients for three crops?	.216*	.191*	.115	.117	1.000	-.447"	.084	.042

main Sig. (bilaterale) inyour environmentN	.027 106	.049 106	.242 106	.232 106	 106	.000 106	.394 106	.672 106
What is the yield correlation coefficient?	.176	-.244*	.167	.147	-.447"	1.000	.292"	-.144
(bilaterale) three cultures	.071	.012	.088	.132	.000		.002	.141
main N corns in the last 5 years in	106	106	106	106	106	106	106	106
your environment (Maiset manioc)								
What is the yield correlation coefficient?	.115	-.038	.153	.029	.084	.292"	1.000	-.077
(bilaterale) three cultures	.239	.700	.118	.764	.394	.002		.434
main N corns over the last 5 years in your area (Beans)	106	106	106	106	106	106	106	106
How does this production correl	-.088	.015	-.049	.093	.042	-.144	-.077	1.000
agricultural for Sig. (bilaterale) crops	.370	.876	.619	.343	.672	.141	.434	
affect N your life socio-economic ue	106	106	106	106	106	106	106	106

**. The correlation is significant at the 0.01 level (bilateral).

*. The correlation is significant at the 0.05 level (bilateral).

DISCUSSION OF THE RESULTS

RESULT 1: Factors contributing to soil infertility in the KAGABI group, particularly in the MBIZA village.

The results of our research showed that the main causes of this problem in this area are, respectively, the lack of improved seeds reported by 28 people out of 106, i.e. 26.4%, the non-rotation of crops reported by 18 people out of 106 interviewed, i.e. 17.0%, non-compliance with ISFM reported by 16 people out of 106, i.e. 15.1%, and the use of traditional soil fertilisation methods reported by 16 people out of 106, i.e. 15.1%.

A comparison of our results with those of KASEREKA LUVAGO (2013) in his dissertation on the "problem of soil infertility in the Mwenga territory: the case of the Basile chiefdom" revealed a similar relationship. The results of his research also revealed that, in this part of the Mwenga territory, soil infertility is at an all-time high as a result of certain factors such as deforestation, the spread of human activities damaging the soil, such as itinerant agriculture over burnt land, mining, the use of traditional techniques, etc. The discrepancy between our study and that of the Basile chiefdom is that it is not possible to determine the extent of soil infertility in the Basile chiefdom. The difference between our study and his lies in the different study environments.

Comparing our results with those of LEBAILLY et al. (2017) when they state that despite the numerous agricultural development plans and programmes that have been formulated in the DRC, and that agriculture has been declared the "priority of priorities" many times already, there is no practical content and no corresponding budget; given this situation, agriculture is regressing at a considerable level and farmland is subject to accelerated degradation, causing infertility.

Lubero Gatumba J. (2015), has shown that ignorance of soil conservation techniques is one of the factors preventing the population in her study area from increasing agricultural production. She showed that erosion and poor cultivation techniques degrade soils, harvests become meagre and food becomes insufficient; all these problems affect aspects of the insufficiency of cultivated land.

In the same vein, NGALAMULUME G (2014) predicts that a good rural development policy should provide the means to meet the challenges facing the countryside while trying to exploit its potential to the full. This should mean maximising the effort and vision for comprehensive rural development in favour of the rural poor by improving returns from working the land, promoting non-agricultural activities and the private sector, setting up innovative partners with the involvement and participation of all rural stakeholders at all levels of action, as well as harmonious sectoral intervention.

Compared with the study carried out in South Kivu by BUCEKUDERHWA C; and MAPENDO S. (2007) in the territory of Kabare, we note a certain similarity with our results as regards the causes of the problem of soil fertility and its management, since these authors came to the conclusion that agriculture in all the territories of South Kivu is faced with several problems, including a reduction in agricultural production, the appearance of new diseases causing the disappearance of certain key crops, the problem of insufficient arable land, and soil infertility, which is 75%. The majority of the population of the province of South Kivu lives below the poverty line, with over 70% of the population living on less than a dollar a day, but the majority of this population also lives in very critical conditions, given that farming activities are not well financed, there are inequalities in the distribution of wealth and income, there are inadequate farm access roads, the soil is degraded and there is little supervisory capacity. But the difference lies in the study environments. These authors focused their research on the entire province of South Kivu, unlike us, who focused on the KAGABI group, particularly in the village of MBIZA.

RESULT 2: Impacts of soil infertility on the socio-economic life of farmers in MBIZA Village ,

To verify our second hypothesis, the field results showed that soil infertility has a negative impact on agricultural yields in the MBIZA VILLAGE, because crops cannot grow and adapt better on infertile soil. 80 people out of 106 surveyed, or 75.5%, said this. Other respondents, 17 out of 106 (16.0%), said that infertile soil had a very negative impact on crop yields.

With this in mind, the FAO (2015) points out that soil degradation is a real and growing threat caused by unsustainable land use and extreme weather events, which are the result of various social, economic and governance factors.

The current rate of soil degradation threatens the ability of future generations to meet their needs, unless this trend is reversed through concerted efforts to achieve sustainable land management. The threat to soils jeopardises the sustainability of agriculture, food security and the provision of ecosystem services. Comparing our results with those of the FAO's research on the impacts of soil infertility around the world, we observe similarities with our results, but the difference is that we have focused on the impacts of soil infertility in a single region, whereas the FAO has generalised these impacts to all regions of the world.

In the same vein, analyses of the degradation of agricultural land and its impact on agricultural production in the town of Kiliba in the territory of Uvira (KIFUVWE MAKI, 2013) showed that agricultural yields in Kiliba remain low for the time being, and this is explained by the fact that 100% of his samples stated that their production had become low as a result of the infertility of the soil. This is explained by the fact that 70% of the surveys stated that the soil is less fertile, especially as a result of poor cultivation techniques in Kiliba.

Comparing our results with those of BUCEKUDERHWA C; and MAPENDO S. (2007) in their study on the causes and consequences of soil infertility in the Kabare territory, reveals a similarity between the negative impacts manifested in our field of study and those noted by these authors in the Kabare territory. Our surveys showed that soil infertility has a negative impact on agricultural yields in the KAGABI group, particularly in the MBIZA village, because crops cannot grow and adapt better on infertile soil, at 75.5%, and BUCEKUDERHWA C ; and MAPENDO have revealed that soil degradation is causing a considerable drop in yields in the Kabare territory because the land is over-exploited, farming activities are not well financed, there is inequality in the distribution of wealth and income, and there are inadequate farm access roads.

RESULT 3: Strategies to be implemented for good soil fertility management and increased agricultural production in MBIZA

The results showed that 27 people out of 106 (25.5%) suggested raising awareness of new cultivation techniques, 20 people out of 106 (18.8%) suggested adopting new crop varieties and 16 people out of 106 (7.5% per response) suggested applying ISFM for good soil fertility management.

By comparing our results with those of TOLLENSE, there is a similarity in the sense that this author proposes that in order to increase agricultural production, we must relaunch and strengthen the public support services for agriculture that the private sector is unable or unwilling to provide, i.e. : agricultural research; agricultural extension and information; agricultural feeder roads; seeds and fertilisers; agricultural credit, strengthening farmers' organisations and structuring the rural world; agricultural training; land security and consultation with the provincial agricultural services. Our surveys also put forward a number of recommendations for increasing agricultural production in MBIZA, which coincide with those of TOLLENSE

1. STRATEGY ANALYSIS MATRIX (PIPO)

A. PROBLEM ANALYSIS (PROBLEM TREE)

Problem analysis is first and foremost about establishing the cause-effect relationships of the negative states of an existing situation. In fact, "the analysis of the problems identified in this work has been in the form of a problem tree", the trunk of which constitutes the central problem, the branches and sub-branches as the effects and, finally, the roots as the causes of the infertility of agricultural soils in the KAGABI group and MBIZA in particular.

Figure 1. Problem tree

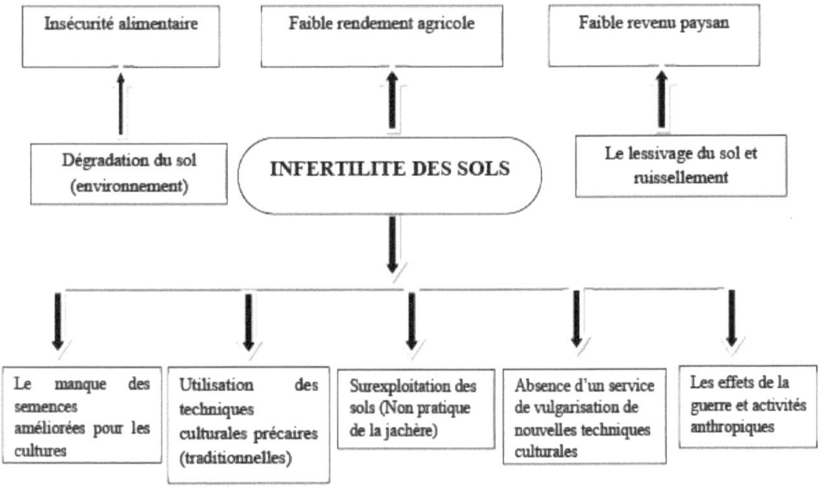

Food insecurity

Low agricultural yield Low farmer income

Soil degradation (environment)

SOIL INFERTILITY

Soil leaching and runoff

Lack of improved seeds for crops

Use of precarious (traditional) cultivation techniques

Overexploitation of the soil (failure to fallow)

Lack of an extension service for new cultivation techniques

The effects of war and human activity

Figure 2. SOLUTION TREE

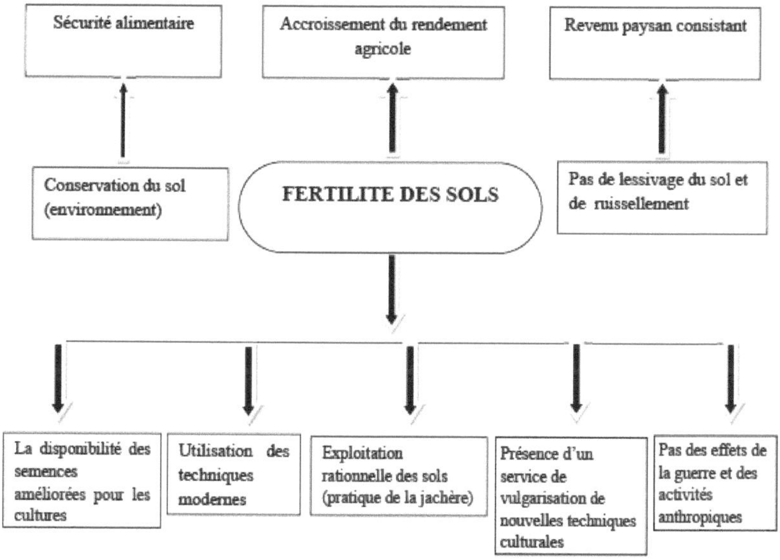

Food safety
Increased agricultural yield
Consistent farm income
Soil conservation (environment)
SOIL FERTILITY
No soil leaching or runoff
Availability of improved seeds for crops
Use of modern techniques
Rational use of soil (fallow land)
Presence of an extension service for new cultivation techniques
Not the effects of war and human activity

Figure 3: Strategy analysis diagram

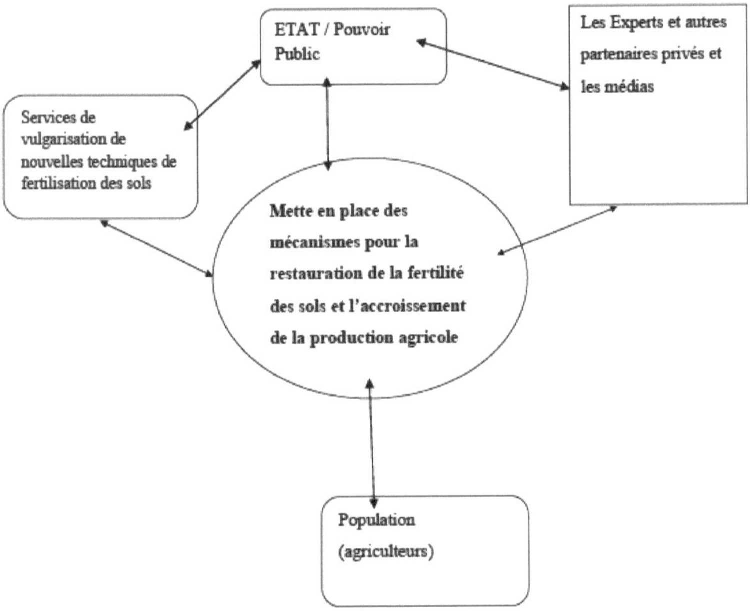

STATE / Public Authority
Experts, other private partners and the media
Put in place mechanisms to restore soil fertility and increase agricultural production
Extension services for new soil fertilisation techniques
Population (farmers)

This diagram illustrates the interaction between the various players who will be working together to solve the problem of soil fertility and its management in agricultural production. A total of 5 players are involved in the success of this strategy, namely the Congolese government, experts, other private partners and the media, the extension service for new soil fertilisation techniques and the general public.

GENERAL CONCLUSION

We conducted research on the "problem of soil infertility and its management in agricultural production in the KAGABI group: the case of MBIZA village". It was found that the problem of soil infertility is at its worst in this area, given that the edaphic situation does not guarantee good soil fertility management and, above all, that the cultivation techniques used are still traditional.

To carry out this work, we used two main materials: the survey questionnaires and the computer.

We used a range of methods and techniques to collect the basic data, including the analytical method, the survey questionnaire, documentation, interviews and participant observation.

At the end of our investigations in the field, we arrived at the following results, which verified our hypotheses:

- The results of our research showed that the main causes of this problem in this area are, respectively, the lack of improved seeds reported by 28 people out of 106, i.e. 26.4%, the non-rotation of crops reported by 18 people out of 106 interviewed, i.e. 17.0%, non-compliance with ISFM reported by 16 people out of 106, i.e. 15.1%, and the use of traditional soil fertilisation methods reported by 16 people out of 106, i.e. 15.1%.
- The field results showed that infertile soil has a negative impact on agricultural yields in MBIZA village, because crops cannot grow and adapt better on infertile soil. 80 people out of 106 surveyed, or 75.5%, said this. Other respondents, 17 out of 106 (16.0%), said that infertile soil had a very negative impact on crop yields.
- The results showed that 27 people out of 106 (25.5%) suggested raising awareness of new cultivation techniques, 20 people out of 106 (18.8%) suggested adopting new crop varieties and 16 people out of 106 (7.5% per response) suggested applying ISFM for good soil fertility management.

RECOMMENDATIONS

- That the Congolese government should take charge of agricultural and food security research centres to enable seed certification and agricultural extension, as well as capacity-building for peasant farmers.
- NGOs working in the agricultural sector should step up their financial support for the agricultural sector in the province of SUD-KIVU in order to revive agriculture;
- Training programmes for farmers on modern cultivation techniques and good soil management should be organised by the Ministry of Agriculture and NGOs;
- Improved inputs should be available in sufficient quantities to benefit the farmers of the SU-KIVU;

BIBLIOGRAPHY

A. WORKS

1. DETRAUX F. et OESTGES, " le *travail du sol et la stabilite de leur* productivite ", edition Vaunbhen, Suede, 2004.
2. FOSTING, La gestion de la fertilite des sols, principes et pistes, ed. Lavauve, Berne, 2002.
3. SEBILLOTTE, *la fertilite des sols a l'horizon 2025 : Quel remede ?* Persee, Paris, 2010.
4. SINGINGA and VERGHEN, *la theorie de Gestion Integree de la fertilite des sols (GIFS),* edition st Laurent, suisse, 2016.
5. TESSIER and ALEX V., Le sol, *une composante essentielle de l'environnement a mieux gerer*: Persee, Paris, 2002.
6. VILAIN; *Rdle agronomique des sols agricoles,* Harmattan, Paris, 2009.

B. REPORTS, REVIEWS AND ARTICLES

7. FAO, *report on soil degradation in the world's regions*, Rome, 2015.
8. IITA; Rapport synthese sur la situation de la détériation des terres cultivables en RDC, Cairo, 2012.
9. LEBAILLY and MPUNGWE, *Plans et programmes de développement agricole en RDC : analyse critique*, article, Loyola, Kinshasa, 2017.
10. BUCEKUDERHWA C; and MAPENDO S., *etude sur la problematique de la baisse de production agricole et ses effets dans le territoire de Kabare,* article, 2007.

C. DISSERTATIONS AND THESES

11 KASEREKA LUVAGO "problematique de l'infertilite du sol dans le territoire de Mwenga : Cas de la chefferie de Basile", dissertation, unpublished, ISDR/ BUKAVU, 2013.

12 . KIFUVWE MAKI, Etat des lieux sur les rendements agricoles a Kiliba et la situation socio-economique des menages, unpublished dissertation, ISDR/ BUKAVU, 2013.

13 LUBERO GATUMBA, *Analyse des facteurs de l'infertilite des sols dans la chefferie des Bahunde en territoire de Masisi* ; memoire inedit, ISDR Grands Lacs/Goma, 2015.

14 COOKE, *soil degradation and poverty in Africa,* article, Abidjan, 2009.

Collecte des données

Source: BAYONGWA BASHIZI David

Data collection
Source: BAYONGWA BASHIZI David

MIX
Papier aus verantwortungsvollen Quellen
Paper from responsible sources
FSC® C105338

Printed by Books on Demand GmbH, Norderstedt / Germany